Pearson's Canal Companion

Cheshire Ring
& South Pennine Ring

Published by Wayzgoose
Tatenhill Common, Staffs DE13 9RS
www.jmpearson.co.uk
email:enquiries@jmpearson.co.uk

WAYZGOOSE

Tenth edition 2013 Updated 2017
ISBN 978 0 9562777 8 7

S PEECH DAY 1967, and the Assembly Hall is rafter-packed with pupils, pedagogues and parents; obscure explorers, B-list bishops, and Old Peterites. Some of the latter - for whom adulthood has proved an anti-climax - are accoutred in brightly-striped blazers, whilst their wives - where they have absentmindedly acquired one - sport Carmen Miranda inspired millinery. At a predetermined signal, the throng shuffle their feet, clear their throats, take their seats, heave a collective sigh, and compose themselves for two hours of speech-making celebrating the achievements of the scholastic year. Significantly, no one has thought to hold a roll call. What pupil in his right mind would dare play truant on Speech Day?

Well, one has, and he is already twenty miles away, changing trains (like Mr Norris) in Castleford. He has never been there before (Castleford, that is, not Berlin) and lacks the clairvoyance to foresee, that over the next fifty years, he will come to know it quite well, and admire it even more; particularly its pies and its barges.

He has gambled on his anonymity. "Pearson keeps an unhealthily low profile," read a recent report to his gregarious father's dismay. "Manifestly not a 'joiner-in', it is not so much that he is ill-behaved, more that he exudes an air of insubordination that masters find disconcerting. On Sports Day - to take just one example - clearly selected to take part in the Intermediate 880 yards race, he was discovered limbering up by the sand pit, claiming that, in front of thirty witnesses, he had been informed he was 'in for the high jump'. The master involved, it transpired, had been speaking metaphorically."

Espousing Abraham Maslow's doctrine of self-actualization, Pearson's grasp of hierarchical needs is advanced for a fifteen year old, and he senses that the journey which lies ahead of him is of more importance than listening to a load of old buffers banging on about the best days of their

Lock-wheeling

lives. Not that he has anything against buffers, let alone best days. His sort of buffers rustily adorn the frames of the Riddles Standard Class 5 wheezing its way into the station in front of his gauzy eyes, one of the last steam-hauled, 'Saturdays Only', services from the Yorkshire coast to the West Riding. A Larkinesque train, 'three-quarters empty'. For already the denizens of Wakefield and Halifax and Bradford are deserting Bridlington and Filey's lugwormed sands in droves, disporting their scantily-dressed, sangria-fuelled alter egos on the Costa-this and Costa-that instead.

'All windows down, all cushions hot, all sense of being in a hurry gone': Pearson requisitions the sort of corner seat Larkin would have elbowed Belloc aside for in a Stanier 'porthole-windowed-corridor-composite'. Normanton ...Wakefield ... Mirfield ... Elland ... Brighouse ... the smokily etched monotones of the still industrialized West Riding roll past the window for his delight and edification, like something from a film by Preston Sturges. This, he inarticulately senses, is the stuff of life. The train bursts forth from a tunnel and crosses a canal. Pearson is, as yet, unaware that this is the Calder & Hebble. Clattering over Greetland Junction, the brakes go on unexpectedly. Thrusting his head out of the window, he is thrilled to see a 'Black Five', joining the rear of the train, to give it a shove uphill to Halifax.

Half a century later, reconnoitering the Calder & Hebble for the purposes of updating this guide, the memory of that occasion comes back to him like an unearthed time capsule. Unwrapping it gingerly, lest its gossamer strands disintegrate, he tautens every sense to see and hear and smell that deep panting train from the past. Wouldn't it be wonderful if every precious memory could be taken down, like a book from the shelf, and fingered lovingly? And have five decades vouchsafed that self-effacing adolescent any wisdom? You bet they have. Never trust *anyone* in Marketing!

Hassall Green

The Cheshire Ring

STUMBLING inadvertently into Staffordshire, the Cheshire Ring picks itself up, dusts itself down, and glances round to see if anyone's noticed this unfortunate *faux pas*, before slipping back unobtrusively into the county implicit in its title. There'll be another indiscretion, when it gets to the outskirts of Manchester. But, for the most part, this 97 mile, 92 lock canal circuit does what its says on the tin, and has established itself as a popular route with boaters, walkers, and - where the towpath's up to it - cyclists as well.

Respectability is regained as the Trent & Mersey Canal passes beneath Pool Lock Aqueduct on the outskirts of Kidsgrove. To reach this ancient, and rather sagging aqueduct from Hardings Wood Junction - the Cheshire Ring's most southerly nodal point - you will have already descended through two locks. Relax, there's only another couple of dozen to go before you reach the bottom of what has been infamously known to generations of boaters as 'Heartbreak Hill'. With the exception of the Pierpoint pair near Hassall Green (Map 2), all the locks were 'duplicated' in the 1830s to speed up the passage of working boats. A further refinement was the provision of paddles between the parallel chambers, enabling one lock to act as a mini-reservoir to

its neighbour. These water-saving side-paddles fell out of use when commercial carrying petered out on the Trent & Mersey in the 1960s, but most of the duplicated chambers remain ostensibly in use, though, with budget constraints, they seem to spend lengthy periods undergoing maintenance.

Twenty-six locks in seven miles between Hardings Wood and Wheelock is not as intimidating as it sounds. Boaters find themselves falling into a rhythm; though, of course, there is no imperative to do it all in one go. Besides, the constituent flights exude their own individual character, and there's scarcely a dull moment as the canal slips off the shackles of urbanisation and meanders through for the most part charming countryside. By Bridge 134, a former warehouse is used by CRT as their Manchester & Pennine Waterways office. The three (out of six) Red Bull locks which follow are photogenically set against a wooded backdrop which masks the Stoke-Crewe railway from view. Up until the railway grouping of 1923, this belonged to the North Staffordshire Railway - affectionately known as 'The Knotty' on account of its heraldic crest incorporating the historic Staffordshire Knot - under whose ownership the canal had been absorbed in 1847. Unlike some railway companies, who

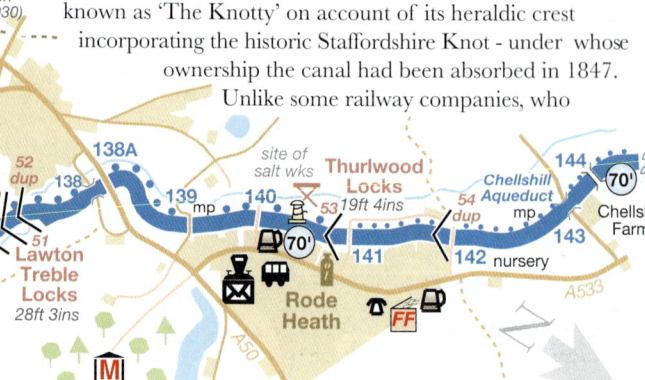

purchased inland waterways in order to suppress them, the Knotty did its best to encourage trade, especially that in raw materials between the Mersey ports and the Potteries. Indeed, up until 1895 they operated their own fleet, after which the Anderton and Mersey Weaver companies carried the bulk of such trade.

Lawton Treble Locks are Telford's work. They replaced a Brindley staircase which was both time consuming and wasteful of water, the former course of the canal lying slightly to the north-east. Bridge 138 (Snapes Aqueduct) carries the canal across the original route of the A50 road whose modern alignment goes over the canal on Bridge 138A. These old trunk roads have been largely rendered irrelevant in the Motorway Age, but they can reward exploration in their own right. This one once linked Northampton with Warrington; an odyssey of acquired taste where most palates are concerned.

In the early years of the *Canal Companions*, Rode Heath boasted two canal associated structures of considerable interest. One was an imposing mill with an arched loading bay. Hearing that the mill, a cherished landmark, was to be demolished, the Trent & Mersey Canal Society successfully applied for the building to be given listed status. In response the mill's owners - doubtless with an eye to the location's development potential - took the matter up with their local MP who had the protected status overturned. Immortally, the DoE stated: 'After further consideration we came to the conclusion that the building was not as interesting as first thought'. The second iconic structure defining Rode Heath's canalscape was the eccentric Thurlwood Steel Lock erected in 1957. Subsidence caused by brine pumping had brought Lock 53 to the brink of collapse, so a new chamber was designed in the form of a steel tank supported on a series of piers which could be raised as the ground around it sank. Entry to the chamber was through guillotine gates. In practise the steel lock took longer to operate than its crumbling conventional neighbour and was mistrusted by boatmen. It had been out of use for a number of years before demolition took place in 1987, but it would have been nice if it could have somehow been kept for posterity as an example of an engineering solution which didn't quite 'cut the mustard'.

Chellshill Aqueduct conveys the canal across a B road to Alsager, an unusually named and curiously self-effacing town two miles to the south. Boaters can take a bit of a breather in the one and a half mile pound between Thurlwood and Pierpoint locks.

Kidsgrove
Map 1

A former colliery town, on the wrong side of Harecastle Hill to qualify as a member of that exclusive hellfire club called The Potteries.

Eating & Drinking
BLUE BELL - Hardingswood (adjacent Lock 41). Tel: 01782 774052. Unspoilt pub prized for an absence of extraneous noise and half a dozen handpumps of ever changing microbrewery ales. Closed Mon. ST7 1EG

RED BULL - adjacent Lock 43. Tel: 01270 782600. One of Robinsons of Stockport's most southerly outposts. Cosily appointed. Food from noon. ST7 3AJ

Shopping
Tesco supermarket, butchers, bakers and branches of the main banks and a launderette. Seek out Kidsgrove Oatcakes on King Street where you can watch oatcakes and pikelets being freshly made on the griddle, and have your oatcakes crammed with a choice of nourishing fillings. Calor gas, diesel and solid fuel are obtainable from Smithsons by Bridge 132. Tel: 01782 787887.

Connections
TRAINS - useful hourly Mon-Sat (and limited Sun) Northern service northwards through Congleton and Macclesfield. Tel: 03457 484950.

Rode Heath
Map 1

Sizeable modern village at junction of A533 and A50. Two pubs vie for your custom: the Broughton Arms (Tel: 01270 878661 - ST7 3RU), canalside by Bridge 140, and the Royal Oak (Tel: 01270 875670 - ST7 3RU), reached from Bridge 142. Both are popular with boaters. There's also a Chinese takeaway called Jade Garden (Tel: 01270 873391), off licence and a well-appointed general store/post office. Buses run to and from the Potteries - Tel: 0871 200 2233.

Things to Do
RODE HALL - 18th century country house and gardens. Tel: 01270 873237. The house and grounds are open to the public on Wednesday and Bank Holiday Monday afternoons throughout the summer. Refreshments. Farmer's Market first Saturday morning of each month (ex January). ST7 3QP

HEAVING a self-congratulatory sigh, the northbound, clockwise boater comes to earth with a bump at the bottom of Wheelock Locks, and offers a wry smile to the upgoing boat crew waiting to replace him or her in the lock chamber: In the paraphrased spirit of the football cliche, they must take each 'Heartbreak Hill' lock as it comes. Not that there's much respite, mark you. Three miles ahead, the locks begin all over again, as the Trent & Mersey makes its way down through the salt-manufacturing town of Middlewich.

Salt? You should have seen this misleadingly bland landscape a hundred years ago! Brine extraction and chemical production had given it the appearance of a First World War battleground. Malkin's Bank (one of the few to emerge with its reputation unimpaired by the scandals of recent years) was traditionally home to the families of boatmen engaged in comparatively short-haul traffics connected with the salt and chemical industries. They lived cheek-by-jowl with employees at the huge Brunner-Mond sodium carbonate works now buried beneath the greens and fairways of a golf course. The works, closed in 1930, was also served by the North Staffordshire Railway's Sandbach branch, sections of which (the Salt Line and Wheelock Rail Trail) have been converted into routes for walking and cycling. Between locks 62 and 63, a side bridge carries the towpath over an old arm (now used by a boatbuilder who appears to specialise in traditional craft) which led into the works.

Hassall Green and Wheelock are both popular lay over points for boaters. At the former, the proximity of the M6 motorway doesn't overly intrude, though perhaps that's because we're all too tired to notice. At the latter, boating facilities are laid on (including recycling) and access is afforded to the not uninteresting town of Sandbach. Several works and mills congregated alongside the canal at Wheelock Wharf and trade here (in corn, fustian, acid et al) was brisk in the heyday of the canal. A small aqueduct carries the canal across the River Wheelock, an insubstantial tributary of the Dane, which flows into the Weaver at Northwich. And the Weaver? Well that, of course, is subsumed into the Mersey somewhere out on Frodsham's lonely marshes.

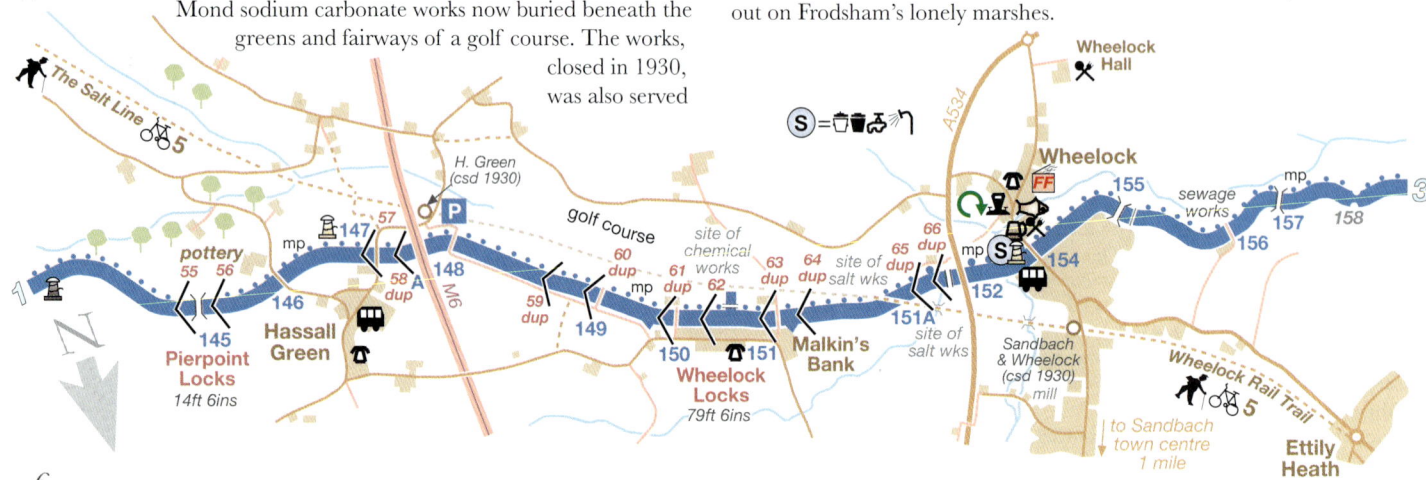

CONCRETE-banked and steel-piled, the canal as it skirts Sandbach, tends to be deeper than normal on account of subsidence caused by salt-mining in the past. The piling is said to be twenty-five feet deep! Sensing this unfamiliar phenomenon, your boat surges forward, relishing the novelty value of clear water beneath its counter.

The canal isn't alone in suffering from subsidence. The Crewe to Manchester railway line has been similarly afflicted as evinced by the huge embankment which carries it in a south-westerly direction from Bridge 158A. This was the site of a transhipment basin between the two modes of transport.

Sandbach Flashes are areas of water brought about by subsidence caused by salt extraction. They form a bird-watching paradise and almost a hundred and fifty species have been identified. No matter that they are located well inland, the flashes are particularly rich in gulls, and even such esoteric species as Glaucous, Icelandic and Caspian gulls have been spotted here, to the rapturous applause of 'twitchers': one man's desultory flock of gulls being another man's nirvana.

Converted into private accommodation in the 1980s, the building alongside Booth Lane Top Lock (No.67) was formerly Moston Mill. It dated from around 1825 and was engaged in grinding corn until the mid-1950s. The other two Booth Lane locks stood alongside what was originally known as Murgatroyd's Salt & Chemical Co. who were in the business of transforming raw salt into various chemicals. Model railway enthusiasts of a certain age will recall that Triang produced an attractive bogie chlorine tank wagon in Murgatroyd's livery in the 1960s. The once extensive works were demolished fairly recently, but a half-timbered farmhouse remains on the site. Dating from the 17th century, it had survived to become the works' sports and social club!

A footbridge above Rumps Lock appears to lead enigmatically nowhere. As with Murgatroyd's, this is the site of another recently vanished industry, the Bisto gravy factory. In contrast British Salt, source of over half the UK's domestic output, remains very much in business.

IN the old days it was salt which brought so much traffic to Middlewich's canals: the salt boats, and, of course, the coal boats, without which, in the pre-electric age, no industry could function. Now, though, it is with pleasure boating that this small mid-Cheshire town is predominantly concerned; two hire fleets and a boatyard and chandlery adding to the often frenetically busy Trent & Mersey Canal which, south of Middlewich, accommodates the orbital traffic of both the Cheshire and Four Counties rings. Sometimes, it seems, there are so many boats wanting to cross your path that you begin to think the town would be more aptly known as *Muddle*wich.

Middlewich Locks form a dog-leg trio, all deep and tediously slow to use. Seddons salt works long ago having been demolished, the canal is rather drearily bounded by light industrial premises, though Middlewich Narrowboats' hire base, with its old canal manager's house and valanced canopy, strikes a welcome element of humanity. Their drydock, on the bend between locks 72 and 73, was once used to maintain Seddons' fleet of narrowboats. The Town Wharf by Bridge 172 handled general cargoes. For many years

its buildings have stood neglected, though in a better setting, one imagines, they'd have been snapped up long ago. Etched interpretive boards summarise Middlewich's history with the emphasis on its Roman origins. Parallel to the tiny River Croco, visitor moorings are provided between Bridge 172 and Big Lock which is named after its width rather than its depth. Adjacent to it stands a redbrick, Dutch-gabled pub of the same name. By Bridge 173 the town's recycling point provides an equally useful facility for boaters.

Croxton Aqueduct was rebuilt to broad-beam dimensions in 1891 so as to permit wide beam craft to work between Anderton and Middlewich. However, after being damaged by flooding in the 1930s, it reverted to its present narrow status. It is proposed that the High Speed 2 railway will cross the Trent & Mersey on a viaduct in the vicinity of bridges 175 and 176. But as it isn't expected to be open until the 2030s, and as civil engineering projects always take longer than imagined, a few more editions of this guide will be published before it needs drawing in on the map. In the woods between bridges 176 and 177 the mangled remains of old wagon tipplers hint at the existence of clay or puddle pits. Note also how the bridges along this length are flat topped so that they could be relatively easily raised in the event of subsidence.

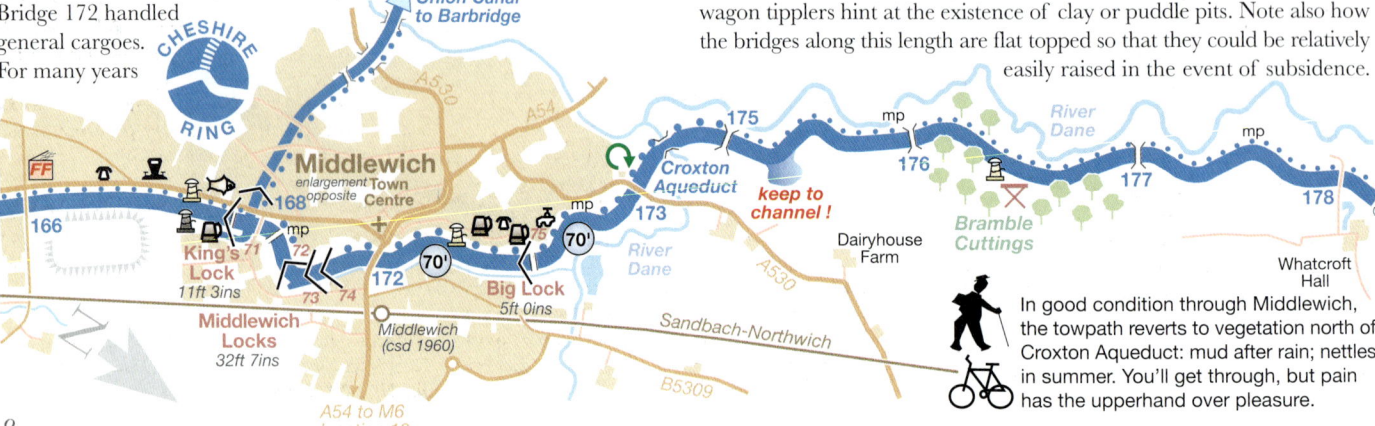

CHESHIRE RING

Shropshire Union Canal to Barbridge

Middlewich
enlargement opposite Town Centre
168

166

King's Lock
11ft 3ins

71
72
73 74

172

Middlewich Locks
32ft 7ins

Middlewich
(csd 1960)

Big Lock
5ft 0ins
70'

70'

70'

173

Croxton Aqueduct

keep to channel !

River Dane

175

176

177

178

River Dane

mp

mp

Dairyhouse Farm

Bramble Cuttings

Whatcroft Hall

Sandbach-Northwich

B5309

A54 to M6 Junction 18

In good condition through Middlewich, the towpath reverts to vegetation north of Croxton Aqueduct: mud after rain; nettles in summer. You'll get through, but pain has the upperhand over pleasure.

Hassall Green   Map 2

Isolated community somewhat impinged upon by the M6, though there are still pleasant walks to be had along the neighbouring country lanes. Downhill, past the shocking pink 'tin tabernacle', the old North Staffordshire Railway has been reinvented as the 'Salt Line' bridleway.

Connections

BUSES - service 78 operates Mon-Sat to/from Rode Heath and Alsager and Sandbach and Nantwich. Tel: 0871 200 2233.

Wheelock   Map 2

Whilst by-passed by the A534, Wheelock still endures more than its fair share of traffic - a culture shock after the peace of the canal. Nevertheless, it's a useful pitstop, and a launch pad for Sandbach.

Eating & Drinking

BARCHETTA - Bridge 154. Tel: 01270 314183. Italian restaurant housed in warehouse conversion, the name roughly translates as 'little boat'. CW11 3RL
CHESHIRE CHEESE - Bridge 154. Tel: 01270 760319. Hydes local. CW11 3RL
Fish & chips (Tel: 01270 768114) and Chinese take-away (Tel: 01270 763169).

Shopping

Convenience store and pet shop.

Things to Do

WHEELOCK HALL - Crewe Road. Tel: 01270 764230. Family run farm shop, tea room (home cooked chips from their own potatoes!) garden centre and play area approximately ten minutes walk south of Bridge 154. Open 9am-6.30pm daily. CW11 4RE

Connections

BUSES - services 37 & 38 provide frequent links with Sandbach in one direction and Crewe in the other.

Sandbach   (Maps 2/3)

Chiefly famous for its ancient Saxon crosses, Sandbach lies about a mile east of the canal at Ettily Heath, though there is easy access to the railway station from Bridge 160. The Town Hall - which houses the covered market, open Thur & Sat - is of decidedly Flemish appearance. In transport circles Sandbach was lauded as the home of lorry making. Fodens had their roots in 19th century agricultural machinery and were at the forefront of the development of steam lorries. Edwin Richard Foden (ERF) broke away from the business to concentrate on diesel lorries and, seeing how successful he became, the family followed suit. Sadly, though, production is no longer centred on the town.

Shopping

Two contrasting outlets make Sandbach worth a detour. Godfrey C. Williams (Tel: 01270 762817) is a 'specialist grocer and cheese connoisseur' located on the cobbled market square. The Beer Emporium (Tel: 01270 760113) on Welles Street offers a cornucopia of bottled beers: British, Belgium and German.

Connections

TRAINS - stopping services between Crewe and Manchester Airport and Manchester. Tel: 0345 7 484950.

Middlewich   Map 4

A salt making town since the days of the Roman occupation, Middlewich's most interesting building is the parish church of St Michael whose tower remains visibly wounded by missiles unleashed during the Civil War - apparently they're still trying to sort out the insurance. Leaflets are obtainable from the library to guide you around some of the known sites of Roman history. June's 'Folk & Boat' Festival continues to grow in popularity and can be relied upon to wake the town from its customary torpor.

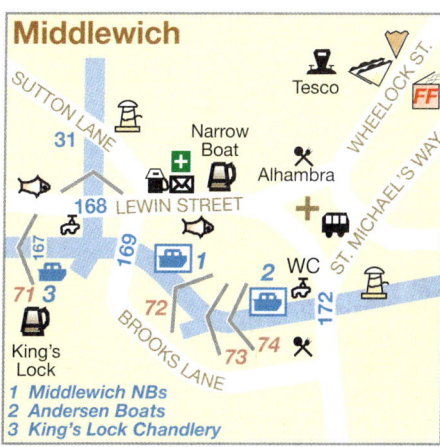

Middlewich

1 Middlewich NBs
2 Andersen Boats
3 King's Lock Chandlery

Eating & Drinking

BIG LOCK - Webb's Lane (Lock 75). Tel: 01606 833489. Food from noon daily. Breakfasts Sat & Sun. Bombardier and guest ales. CW10 9DN
KING'S LOCK INN - Booth Lane (Lock 71). Tel: 01606 836894. *Good Beer Guide* listed canalside pub highly thought of for its food. CW10 0JJ
THE NARROWBOAT - Lewin Street. Tel: 01606 738087. Welcoming town centre pub with a well-appointed dining room. CW10 9AS

Shopping

There are Lidl, Morrisons and Tesco supermarkets, NatWest and Barclays banks, and a small market is held every Tuesday. Local produce from Middlewich Narrowboats canal shop.

Connections

BUSES - Arriva service 42 links Middlewich with Congleton and Crewe (for the railway station). Tel: 0871 200 2233.

WOODLAND interludes and subsidence-induced flashes characterise the Trent & Mersey's serene passage through the Dane Valley. Hereabouts the river (having risen in the Derbyshire Peak District on the flank of Axe Edge - and having been crossed again by Cheshire Ring travellers at Bosley - Map 21) has grown sluggish with age, meandering about its level valley in a series of lazy loops; one moment it is hard by the canal, the next away across the pasturelands of milking herds. The soil here is soft and the Dane carves deep banks made shadowy by alder and willow. The canal shares the valley with a Roman Road known as King Street and a now lightly used railway which once sported a 'push & pull' service between Crewe and Northwich. But these other transport modes barely intrude upon what is otherwise a long, relaxing pound. At Higher Shurlach commerce rears its perfidious head in the shape of a giant distribution depot for Morrisons. The moist aromas of bread-making issue from Roberts 'Red Rose' bakery who boast their own brass band. Altogether now - dough, re, mi ...

The most curious feature of this section of the canal are the flashes bordering the main channel to the south of Bridge 181. That nearest the bridge was once filled with the submerged wrecks of abandoned narrowboats, an inland waterway equivalent of Scapa Flow. Many of the boats were brought here and sunk *en masse* during the Fifties in circumstances as controversial - in canal terms that is - as the scuttling of the German Fleet after the First World War. In what was probably a book-keeping exercise, British Waterways rid themselves of surplus narrowboats in a number of watery graves throughout the system. In recent years the wrecks have been raised and taken off for restoration. You know how it works! One generation's cast-offs become the next's prized possessions.

The Tata works at Lostock dates from the late 19th century. Pipes spluttering with steam span the canal and lime waste is pumped across to the lagoons hidden by high embankments beyond the A530. The works, previously owned by Brunner Mond and ICI, retains a handsome brick-built, Dutch-gabled office block, but has lost the canal basin which once served it, the only remaining clue being a side bridge which lifts the towpath over its former entrance by the railway.

West of the canal lies the Help For Heroes stadium of Witton Albion, founded in 1887. From 2005 to 2012 this proud club could look directly across the canal to the stadium of an even older football club, Northwich Victoria, who could trace their origins back to the early 1870s. 'The Vics' had played at the Drill Field for a record hundred and twenty five years before moving to their new purpose built, canalside stadium in 2005. Formally opened by Sir Alex Ferguson, so impressive did this new Victoria Stadium look that Nike used it in 2008 to film an advert directed by Guy Ritchie (husband, you'll recall, of Madonna) and starring a stellar cast of footballers including Wayne Rooney, Cristiano Ronaldo and Carlos Tevez. Four years later the stadium, barely seven years old, was demolished to facilitate the expansion of a neighbouring chemical plant, and Northwich Victoria are now reduced to playing their home games on the opposite bank of the canal. None of the foregoing has been made up. You couldn't!

Broken Cross Map 5

Canalside at Bridge 184 stands the Old Broken Cross refurbished in the modern manner - Tel: 01606 40431. CW9 7EB. Garage with Spar shop less than five minutes walk to south-east. Co-op ten minutes west.

Wincham Wharf Map 5
Eating & Drinking
NECTARS BAR - Manchester Road (Bridge 189). Tel: 01606 333723. Warehouse conversion. CW9 7NT
THE CODFATHER - Manchester Road. Tel: 01606 42342. Excellent fish & chips. CW9 7NE
MR FU's - Manchester Road. Tel: 01606 42246. Chinese buffet. CW9 7NE

Marston Map 5
Former salt mining village still wrought by the lop-sided scars of the past.
Eating & Drinking
SALT BARGE - Ollershaw Lane (Bridge 193). Tel: 01606 43064. CW9 6ES
Things to Do
LION SALT WORKS - Ollershaw Lane (Bridge 193). Tel: 01606 275066. Award-winning heritage centre devoted to the history of salt-making. Open Tue-Sun 10.30am-5pm. Gift shop, cafe, play area. CW9 6ES

Northwich (Map 5)
A fascinating - one might even go as far as saying endearing - town located just off the map on the Weaver Navigation. Of course the best way of getting there is down the Anderon Lift by boat, a route fully

Anderton Lift

Robin Smithett

covered in Pearson's *Four Counties Ring Canal Companion*. Motives for visiting Northwich range from its excellent Salt Museum to a Webb's pork pie. In the absence of a boat, buses run from Anderton or Barnton (Map 6) - Tel: 0871 200 2233.

Anderton Map 6
Good moorings, the resonating proximity of 'The Lift', and a country park with a Wildflower Trail, render Anderton a popular stop in many an itinerary.
Eating & Drinking
THE MOORINGS - restaurant and coffee shop at Anderton Marina. Tel: 01606 79789. Open daily (ex Tue) 11am-5pm and 6.30-9pm. CW9 6AJ
STANLEY ARMS - Old Road (canalside opposite Anderton Lift). Tel: 01606 75059. CW9 6AG
LIFT CAFETERIA - canalside. Visitor centre cafe offering fine views of the Lift.
Things to Do
ANDERTON BOAT LIFT - Tel: 01606 786777. Canalside Visitor Centre celebrating the Lift and local canals in all their historic glory. A widebeam trip boat named *Edwin Clark* after the Lift's designer offers trips up or down the Lift. River trips to Northwich and back are also usually available. CW9 6FW
ANDERTON NATURE PARK - waymarked trails through reclaimed wasteland where many plants usually confined to coastal environments thrive in the local salty soils.
Connections
BUSES - service 46 runs to/from Northwich approx bi-hourly Mon-Sat. Tel: 0871 200 2233.
TAXIS - Northwich Taxis. Tel: 01606 46666.

6 TRENT & MERSEY CANAL Anderton & Barnton 5mls/0lks/2hrs

EXCITEMENT mounts, whatever your mode and direction of travel, as the canal nears Anderton and its famous Boat Lift. Of all the so-called "Seven Wonders of the Waterways", Anderton Lift is arguably the most ingenious, and it was a relief to have it back in our midst, operating again as a link between the Trent & Mersey Canal and Weaver Navigation* after twenty years in the disused and rusting wilderness between 1983 and 2002. If you aren't planning to use it, at least pause awhile and watch it being put through its paces. Down on the banks of the Weaver a popular visitor centre supplies the facts and figures behind this Heath Robinson like contraption. Suffice it here to say that it dates from 1875 and was designed by Edwin Clark who, *Canal Companion* aficionados will pretty much uniquely know, hailed from Marlow; and who, some three years earlier, had engineered the railway to that Thameside town.

Either side of Anderton, the Trent & Mersey Canal continues in its unhurried, Brindleyesque manner, negotiating a

scarred landscape destabilised over the years by salt extraction. In 1958 a new length of canal had to be dug at Marston to bypass a section bedevilled by subsidence, but such is nature's capacity for reclaiming her own, you'd be hard pressed to discern many signs of former industry along these bosky lengths of the canal.

Barnton and Saltersford tunnels were amongst the earliest essays in the art of canal tunnel digging. They are far from straight. Neither are they wide enough for narrowboats to pass inside. Their lack of width also prevented widebeam barges from traversing the Trent & Mersey between Preston Brook and Middlewich as had originally been planned. They are short enough, however, not to cause undue traffic delays. Both bores are towpathless, but walkers have the bonus of a charming walk, by way of the old horse paths, across the wooded tops. Separating the two tunnels is an idyllic, leafy pool, which provides a pleasant mooring. You could linger here indefinitely, watching the boats emerge from the tunnels, and descending to the Weaver for a stroll along the riverside path.

turn back to page 11 for details of facilities at Anderton

*Coverage of the Weaver Navigation appears in Pearson's Four Counties Ring Canal Companion.

⚠ Northbound boats may enter Saltersford Tunnel for 20 minutes on the hour, southbound similarly on the half hour.

Carey Park · Furey Wood · Wallerscote Works (dis) · Winnington Works · Anderton Boat Lift · Uplands Basin Marina · Anderton Nature Park · 70' · 196 · 198 · 199 · 200 · 201 · Anderton · Anderton Marina ABC Boat Hire · Marbury Country Park · Marston New Cut · Barnton Wharf · Barnton Tunnel 572 yards · Barnton · Saltersford Tunnel 424 yards · Saltersford Lock · Weaver Navigation · 204 · 205 · 206 · 7 · A533 · Little Leigh

PRESTON BROOK is one of those nodal points on the inland waterway follower's emotional compass that none of their more rationally grounded friends and acquaintances will have ever heard of. Here - just inside the northern portal of Preston Brook Tunnel - that doyen of the canals, the Bridgewater, meets the Trent & Mersey. In the heyday of water transport Preston Brook was one of the busiest canal centres in the north-west. An inland port where cargoes were transhipped between wide-beam, Mersey 'flats' and narrowboats. A substantial number of warehouses were erected to cater for this activity, which continued up until the end of the second world war. Indeed, narrowboats continued to trade here until the late 1960s, following which the majority of warehouses were

regrettably demolished. And, consequentially, Preston Brook, whatever its historic status, lacks the obvious appeal of peers such as Shardlow or Stourport. A notable survivor, however, is a former flour warehouse now converted into flats, whilst there are plenty of boats usually to be seen on the move; a long-established hire base and a marina see to that.

North of the M56 motorway - which connects Manchester with North Wales - the Bridgewater Canal rapidly establishes a rural atmosphere. On the western horizon the cooling towers of Fiddlers Ferry power station form a notable landmark, though the plant was due to cease generating electricity in 2017. Norton's prominent water tower was built in 1892 as part of the system which supplies Liverpool with water from Lake Vyrnwy in mid-Wales. Closer to hand, the village of Daresbury has connections with Lewis Carroll, but is these days perhaps better known as home to a world renowned Science & Innovation Campus dominated by a lofty tower housing a Van de Graaff accelerator.

Map labels:

8A · water tower · Norton · Runcorn East · Norton Town · Murdishaw · Cawley's Bridge · Borrow's · P. Brook Marina · viaducts · A558 to Runcorn · M56 to N. Wales · A56 to Chester · Red Brow Underbridge · 82 · George Gleaves · Crows Nest Farm · Keckwick · Moore · rugby club · Preston Brook · P. Brook - csd 1948 · Claymoore M. Chandlers · warehouse · Daresbury Park · Crow's Nest · Daresbury Firs · Keckwick Hill · Science & Innovation Campus · twr · Moorfield · 70' · Moore · 55' · mp · Preston on the Hill · A56 · 11 · Daresbury · 1 Daresbury - closed 1952 · 2 Moore - closed 1943 · Acton Grange · Preston Brook Tunnel · 1239 yards · CHESHIRE RING · A56 to Warrington · Thomason's · 9 · MANCHESTER SHIP CANAL

Designated the Bridgewater Way, the Bridgewater Canal's towpath is earmarked for upgrading, though progress is reliant on funding packages and therefore tediously slow. Cycling (seemingly equated with dog-fouling) is currently prohibted on unimproved sections.

8A BRIDGEWATER CANAL Runcorn 3mls/0lks/1hr

ANYONE with an
affinity with the flotsam
and jetsam of small coastal
ports is likely to find a detour off
the main route of the Cheshire Ring into
Runcorn an invigorating experience. Less perceptive observers have
been flippantly dismissive with the arm's five lockless and predominantly
urbanised miles, and visiting boats are a relatively rare phenomenon.
Indeed, most vessels on the move seemingly belong to members of the
Bridgewater Motor Boat Club which has premises in the old Sprinch
boatbuilding yards at the western end of the canal.

But if boats are rare, towpath walking and cycling are activities
enthusiastically embraced by the native population, as is angling. And
there are rural interludes of a sort: notably in the vicinity of Greens
Bridge, where rhododendron shrubs fill the woods bordering the lily-
fringed canal, and a bosky mooring is provided for boaters wishing to
visit Norton Priory. Here too lies the northern edge of Runcorn's Town
Park, which boasts an extensive miniature railway that operates on
Sunday afternoons throughout the year. West of Norton views open out
over the Manchester Ship Canal and the widening river with its new
Mersey Gateway toll bridge. This length used to be bordered by
tanneries - producing leather for shoes and suitcases, harnesses and
handbags - and soap works. During the Great War mustard gas was

for details of facilities turn to page 18

produced on Wigg Island, now it slumbers as a nature reserve. Here it was
that the Guinness boats use to dock, bringing vital supplies of Dublin's finest.
Under the flight path for John Lennon Airport, the canal approaches its
foreshortened terminus, passing Co-operative Society premises, funeral
parlours and pubs heavily shuttered out of hours. Stumps of old cranes hint
that business was once brisk, but the arm meets its Waterloo, so to speak,
at an ornately eponymous bridge which used to span the top chambers of
the much-mourned Runcorn Locks. From here two flights of ten locks each
- the Old and New locks - led down to link with the Manchester Ship and
Runcorn & Weston canals.

There are long term proposals that the Old Locks might one day be
restored. In the meantime their course (and some obvious remains of former
chambers) can be followed by pedestrians, whose curiosity will be rewarded,
at the foot of the flight, by Bridgewater House, the 'Canal Duke's' residence
of 1777, erected so that he might personally oversee completion of his
canal. Runcorn Docks are still in business, and receive fertilizers, bulk clay
and other commodities by ship.

17

Bartington Wharf — Map 7

Canalside community on the busy A49.

Eating & Drinking

LEIGH ARMS - Tel: 01606 853327. Comfortable Robinson's outpost overlooking Acton Swing Bridge. Painted windows featuring local scenes. CW8 4QT
THE HOLLY BUSH - Tel: 01606 853196. 17th Century thatched pub on A49. Food lunch and evenings (from 5.30pm) Mon-Sat and from noon Sun. CW8 4QY.

Shopping

DAVENPORT'S FARM SHOP - Tel: 01606 853241. Award-winning farm shop, florist and tea room open daily 10am-4pm ex Tue. CW8 4QU

Connections

BUSES - service 48 runs bi-hourly (ex Sun) to/from Northwich and Frodsham. Tel: 0871 200 2233.

Dutton — Map 7

BLUEBELL COTTAGE GARDENS can be found just downhill from Bridge 213. Tel: 01928 713718. The garden showcases a range of herbaceous hardy perennials available to purchase from the adjoining organic nursery. Additionally there are bluebell woods and a wildflower meadow to explore. Light refreshments. WA4 4HP

Preston Brook — Map 8

Spar convenience store with post office counter just a couple of minutes walk west of canal. Buses to Runcorn.

Daresbury — Map 8

A bracing ten minute walk uphill from Keckwick Bridge leads across the A56 into the peaceful core of Daresbury (pronounced 'Dars - as in stars - bury') village. Uplifted by a visit to All Saints parish church (Lewis Carroll souvenirs obtainable) and having viewed the handsome Sessions House, you can justifiably seek out refreshment at the Ring o' Bells (Tel: 01925 740256 - WA4 4AJ) a Chef & Brewer pub which serves food from 11am daily (noon on Sundays). The X30 bus links Daresbury with Chester, Runcorn and Warrington. Tel: 0871 200 2233.

Moore — Map 8

Peaceful village despite its proximity to Runcorn. Home of the charmingly named Gentlemen of Moore Rugby Club. Good pub called the Red Lion (Tel: 01925 740205 - WA4 6UD), village stores and post office adjoining canal, and buses to/from Warrington and Runcorn. Moore Nature Reserve can be reached by walking down Moore Lane and crossing the Manchester Ship Canal by way of one of its characteristic swing bridges. The reserve, consisting of two hundred acres of woodland, meadows, lakes and ponds, intriguingly incorporates an abandoned length of the old Runcorn & Latchford Canal.

Runcorn — Map 8A

A former dockyard town with a Milton Keynes-like expansion grafted on. Aesthetically, some of the new town works, some of it doesn't, but the old town makes up for any such failures with the excitement of its setting overlooking the magnificent width of the river. The huge suspension bridge dates from 1962. It replaced a transporter bridge which could carry only four road vehicles at a time. Stanley Holloway immortalised it in his comic monologue *Tuppence Per Person Per Trip*. A bracing walk across its haughty replacement can be heartily recommended, perhaps coupled with a visit to Catalyst (Tel: 0151 420 1121) the science discovery centre at West Bank on the far side. Or one might perhaps go and pay homage at Widnes railway station to the plaque which records that Paul Simon wrote *Homeward Bound* on its lugubrious platform.

Eating & Drinking

THE BRINDLEY - Tel: 0151 907 8360. Canalside arts centre with bar and terrace cafe. WA7 1BG
THE FERRY BOAT - Church Street. Tel: 01928 583380. Wetherspoon with nice illustrations of Runcorn's maritime past. WA7 1LR

Shopping

Enough shops remain in the town centre to meet canallers' needs. Tucked away in Granville Street (opposite the library) is Monk's delicatessen, excellent for both cooking ingredients and take-away food. Staunchly independent, Curiosity Books (Tel: 01928 575956) are on High Street opposite Lloyds Bank. The Market Hall is open daily and there's a street market on Tuesday. The post office is opposite the bus station.

Things to Do

NORTON PRIORY - Warrington Road. Tel: 01928 569895. Canal access from Green's Bridge. Open daily from 11am-4pm. Three distinct layers of history which peel back to reveal: life in the original monastery dissolved by Henry VIII in 1536; the Tudor mansion which grew up in its place; and the Georgian country house of 1740 that succeeded it. WA7 1SX

Connections

BUSES - an unusual feature of the town is its segregated 'busways' which provide a fast 'interurban' network worth sampling for its own sake. Try the No.3 bus on its circular run via Weston village for exhilarating views over the Mersey, Weaver and Ship Canal. Tel: 0871 200 2233.
TRAINS - station adjacent to canal terminus well served by trains to Liverpool, Crewe, and London Euston. Runcorn *East* station is on the Manchester, Warrington, Chester and North Wales route and is handily placed for Preston Brook. Tel: 03457 484950.
TAXIS - Apec. Tel: 01928 575757.

9 BRIDGEWATER CANAL Stockton Heath 4mls/0lks/1.5hrs

WARRINGTON - one of three towns in Britain with a transporter bridge, but unique in having one which doesn't work - lies close to the Bridgewater Canal, but you wouldn't know it, the Manchester Ship Canal and the interminable suburbs of Stockton Heath, keeping it out of sight and - somewhat undeservedly - out of mind. But Stockton Heath is a good spot for a breather, irrespective of whether you're afoot or afloat. For besides its facilities - including a cosmopolitan array of eating establishments - the opportunity arises to stroll down London Road and inspect the Ship Canal, marvelling at the advances accrued in a century and a quarter of canal building between completion of the Bridgewater Canal in 1770 and the MSC in 1894. And, if you're luckier than us, you might see the swing-bridges being swung and the stately progress of a sea-going vessel, notably the *RMS Veritas* which runs from Seaforth (Liverpool) to Irlam four times a week with containers, a voyage featured on Channel 4's programme

Rivers, presented by Jeremy Paxman.

West of Stockton Heath, the Bridgewater Canal journeys through charming countryside, negotiating a lush cutting ripe with rhododendrons at Higher Walton, a leafy estate village built by Victorian brewing magnates and featuring a Gothic Revival church by Paley & Austin. From Hough's Bridge a by-road leads southwards to Appleton Reservoir. Whilst, from the neighbouring escarpment, there are fine views over the Mersey Valley.

At London Bridge, the premises of Thorn Marine were originally a Bank Rider's house with stabling for boat horses. The bridge itself dates from the 1930s when London Road was upgraded. On the south-eastern side of the bridge, by the London Bridge pub, look out for the semi-circular flight of steps down to the water's edge. This was Stockton Quay where 18th century canal passengers would embark aboard packet boats linking Runcorn with Manchester.

Soaring above suburban rooftops, the canal crosses Lumbrook Under-bridge (as aqueducts on the Bridgewater Canal are known) and winds past the village of Grappenhall.

for details of facilities at Higher Walton, Stockton Bridge and Grappenhall turn to page 21 *19*

10 BRIDGEWATER CANAL Lymm & Thelwall 4mls/0lks/1.5hrs

AN entertaining walk is there for the taking at Thelwall. Go down the steps from the underbridge and pass beneath the old railway - they used to win prizes for their flower displays at Thelwall station before it closed in 1956 - cross the A56 and proceed down Bell Lane, bearing left by the village cross. On reaching the 'Pickering's Arms' pub (note the inscription commemorating the foundation of a settlement here - 'city' being something of an exaggeration - in 923 by the son of King Alfred the cake burner) turn right down Ferry Lane. Regulars will know of the enchantment ferries hold for us, and the Thelwall Ferry (operating hours 7-9am ex Sun, 12-2pm and 4-6pm) has always been one of our Cheshire Ring treats. The ferryman will 'scull' you over to the far bank of the Manchester Ship Canal in his aluminium boat. Having gained the far bank, a short walk westwards (along what used to be the course of the MSC's own internal railway system) will bring you to the giant Latchford Locks (Map 9). The larger of the two chambers measures 600 x 65ft - just think how many narrowboats you could fit in that. You can complete the circle by turning right at the Spar shop and going up on to the Trans Pennine Trail, following the old railway back to Thelwall, mildly cussing that the nanny state denies access to the lofty girder bridge which once carried the trains across the Ship Canal. Incidentally, the ferry fare is currently astonishing value at just thirteen pence, but in the first edition of this guide it was only half a decimal penny. Sadly we're incapable of calculating what rate of inflation that represents, though doubtless some of our more numerate users will know!

Unostentatiously true to its 83ft contour, the Bridgewater Canal slips through the dainty little town of Lymm. In canal terms, this twenty-five mile pound constitutes what the French would call the *longueurs* of the Cheshire Ring. So one welcomes the charm of Lymm and is grateful for an excuse to leave the canal behind and renew one's relationship with dry land. The inconspicuous Whitbarrow Aqueduct carries the canal over a brook which cleaves

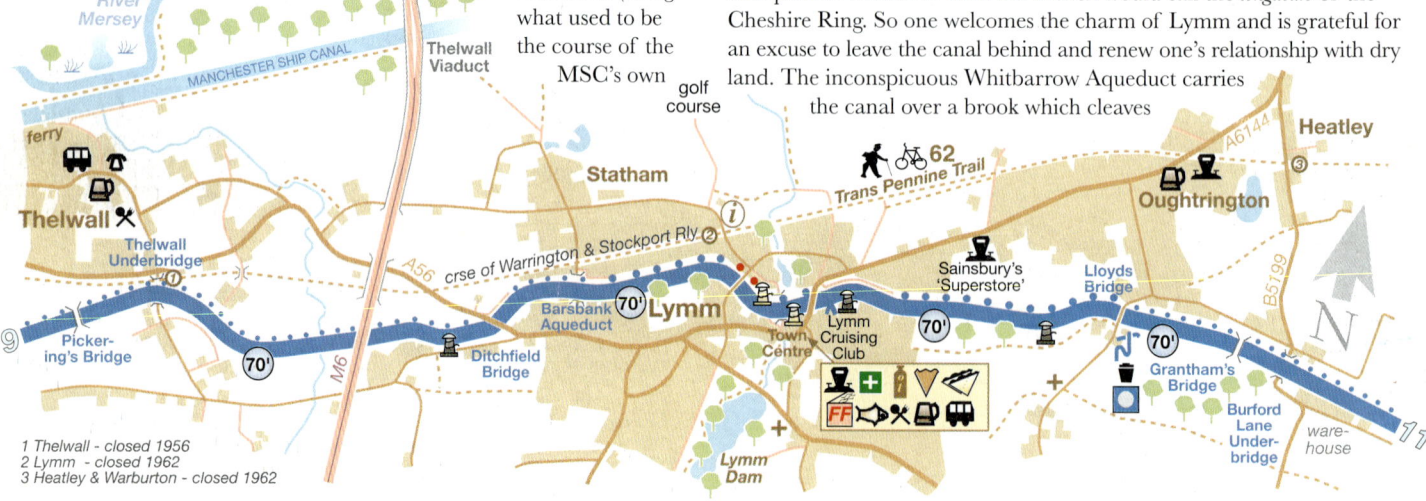

1 Thelwall - closed 1956
2 Lymm - closed 1962
3 Heatley & Warburton - closed 1962

a dramatic gorge through a sandstone outcrop on its way down to the Mersey. The gorge has been turned into an attractive park, but once there was a mill here, the sluice and water wheel of which remain to tickle your historic fancy. Lymm (opponents of HS2 take note) was an early victim of transport engineering. Brindley surveyed a route which sliced the old town square in half, so that even today houses overhang Lymm Bridge like interrupted conversations.

South of Lloyds Bridge (where useful facilities for boaters are obtainable) Oughtrington's parish church is seemingly displaced from the rest of the village. In common with Walton's (Map 9) it was built from Victorian

profits, in this instance by one of the 'cotton kings', George Charnley Dewhurst, who resided in Lymm. To the north of Lloyds Bridge stands a three-storeyed terrace of former fustian-cutting cottages, the raw cloth being brought in from Manchester by canal boat, treated and returned as velvet. A fine example of a canalside warehouse stands to the east of Burford Lane Underbridge. Note how there is no banking, the canal water lapping its brickwork, and rendering, one assumes, the ground floor prone to damp. The adjoining house has been beautifully restored but the warehouse has patently seen better days. In the 1980s it was used as a workshop producing spare parts for the iconic Vincent motorbike.

Higher Walton · Map 9

Estate village built on the brewing fortunes of the Greenall family. Walton Hall is a wedding venue, but the grounds are a popular destination for Warrington townsfolk, not least the rose garden, woodland walks, bandstand, bowling green, pitch & putt, heritage centre and children's zoo. Close to the canal stands the Walton Arms (Tel: 01925 262659 - WA4 6TG) a popular carvery/pub.

Stockton Heath · Map 9

An unexpectedly prosperous suburb of Warrington where you can eat Chinese, French, Indian, Italian, Thai or Turkish, though there are pubs and fish & chips as well for unreconstructed palates. Plenty of shops too, ranging from a Sainsbury's 'Local' just down from London Bridge to a large Morrisons supermarket across the Ship Canal. Frequent buses run into Warrington, a likeable town with an admirable indoor market.

Grappenhall · Map 9

A pretty cobbled lane leads to the 16th century church and village stocks. There are two nice pubs on Church Lane: Rams Head (Tel: 01925 269320) and Parr Arms (Tel: 01925 212120.) WA4 3EP

Thelwall · Map 10

Leafy community overlooking the Ship Canal. Little Manor (Tel: 01925 212070 - WA4 2SX) is a country pub and restaurant open from noon daily, whilst the Pickering's Arms (Tel: 01925 861262 - WA4 2SU) is a half-timbered pub close to the ferry.

Lymm · Map 10

Disingenuously billed as a 'village' but, with a population of eleven thousand, more like a town, Lymm is nevertheless one of the most appealing ports of call on the 'Cheshire Ring', and its facilities could not be handier. An enigmatic sandstone cross occupies the centre, and there are several picturesque nooks and crannies. Lymm Dam dates from the construction of the Stockport to Warrington turnpike in 1824.

Eating & Drinking

BREWERY TAP - Bridgewater St. Tel: 01925 755451. *Good Beer Guide* listed microbrewery. WA13 0AB
ELMAS - Pepper Street. Tel: 01925 756049. Mediterranean restaurant and take-away. WA13 0JB
FLAVOURS - The Cross. Tel: 01925 753079. Turkish restaurant. WA13 0HU
RAYMONDO'S - The Cross. Tel: 01925 756067. Italian restaurant. WA13 0HU

SEXTONS - Eagle Brow. Tel: 01925 753669. Cafe serving fresh baking. WA13 0AD
There are several more pubs within easy reach of the canal with opportunities for connoisseurs to sample Manchester brewed beers by Hydes and Lees. Also near at hand are take-aways and fish & chips.

Shopping

Conscious of its charm, boutique and antique outlets have infiltrated Lymm's more traditional retailers, amongst which are Hopkinsons butchers, Sextons bakery (which does take-away/ready cooked meals), pharmacy, and wine merchant. Sainsbury's boast both 'Local' and 'Superstore' outlets. Small canalside market on Thursdays. Food & craft market 3rd Sunday in the month.

Things to Do

LYMM RANGER CABIN - Tel: 01925 758195. Situated on the Trans Pennine Trail and starting point of an entertaining waymarked Heritage Trail which is a great way of getting to know Lymm. WA13 9NJ

Connections

BUSES - services 5, 37 & 38 offer an approximately quarter-hourly (hourly Sun) link to/from Warrington and Altrincham. Tel: 0871 200 2233.

OUTSKIRTS exude their own remoteness; a quality exaggerated by the sense of contrast between what is urban and what is rural. Greater Manchester's sprawl hasn't quite extended to its boundary with Cheshire. Oldfield Brow marks the beginning (or end in the case of westbound travellers) of the conurbation. Dunham Town sits prettily in the fragility of no-man's-land. From hereabouts to Romiley - the first semblance of countryside on the far side of Manchester - it is ten hours cruising. Long gone, though, are the days when we would have exhorted you to avoid mooring overnight in the city.

The Bridgewater Canal is not as shallow as most: clearly a case, as Cat Stevens so perceptively put it, that "the first cut is the deepest." But long lines of moored craft force you to throttle down to a sedate pace, so the opportunity to accelerate without a wash must be postponed. Why hurry? The flat, rural landscape has its own refreshing charm; peaceful and timeless being epithets which spring easily to mind.

Coal used to be unloaded at Bollington Wharf. On the adjacent road, the Olde No.3 pub gained its name from the fact that it was the third stop on the coaching route

from Liverpool to London. Of a shy and solitary nature (a default-setting disposition where many canal enthusiasts are concerned), Maurice Egerton, the 4th and last Baron Egerton of Tatton Park, continued his forebears enthusiasm for canals by keeping an ex-army launch here in the 1950s. Maurice was descended from Francis Egerton, the third Duke of Bridgewater, who had so famously promoted the Bridgewater Canal, whilst his uncle, Wilbraham Egerton, was Chairman of the Manchester Ship Canal, cutting the first sod in 1887. A great traveller, pioneer motorist and aviator, Maurice left the family's seat at Tatton Park (4 miles to the southeast) to the National Trust, dying on his Kenyan estate in 1958.

There are pleasant views across parkland roamed by fallow deer to Dunham Massey's 18th century mansion, open to the public under the aegis of the National Trust. Towards the end of the Second World War the estate was used as a prisoner of war camp whose inmates erected a large model of a Bavarian castle made from scrap materials. Post-dating the estate, the canal slices across an avenue leading from the house to an obelisk in an oak wood which was erected to commemorate a racehorse called Bay Malton, though there is no inscription as such. The horse, which anecdotally won enough prize-money to save the owner of Dunham Massey's fiscal bacon, was also remembered by the name of a canalside pub at at Oldfield Brow.

Between Warrington (Map 9) and Broadheath (Map 12) the canal often keeps company with the Trans Pennine Trail, a 215 mile multi-user route from Southport to Hornsea, hereabouts occupying the course of the former Warrington & Stockport railway, closed to passenger trains as long ago as 1962, but used by freight (mostly Yorkshire coal bound for Fiddler's Ferry power station) until 1985.

Skirting Altrincham, the canal encounters the Linotype works for manufacturing letterpress type. In these digital times it seems inconceivable that print was produced until quite recently with lumps of metal, but at its zenith the works - whose imposing clocktower stands at right-angles to the canal - gainfully employed a couple of thousand souls whose descendants probably now work in call centres.

Agden Wharf   Map 11
BARN OWL - Warrington Lane. Tel: 01925 752020. Canalside pub, sadly no longer operating the ferry which once provided access for thirsty towpathers. Food, Thwaites and guests. WA13 0SW

The Dunhams   Map 11
Picturesque estate villages, Woodhouses and Town, lie on either bank of the canal and repay exploration. St Marks, at Dunham Town, is a quaint little Victorian church dating from 1865 with Stamford tombs.

Eating & Drinking
THE OLDE NO.3 - Lymm Road (A56). Tel: 01925 756754. Roadside pub adjoining off-side visitor moorings. WA14 4TA
THE SWAN WITH TWO NICKS - Park Lane. Tel: 0161 928 2914. A classic country pub recommended in the *Good Beer Guide*, featuring locally-brewed Dunham Massey ales amongst others, and a wide range of food. WA14 4TJ
AXE & CLEAVER - School Lane. Tel: 0161 928 3391. Chef & Brewer pub/restaurant. WA14 4SE
VINE INN - Barns Lane, Dunham Woodhouses. Tel: 0161 928 3275. Quaint country pub offering an all too rare opportunity to sample Sam Smith's ales from Tadcaster. WA14 5RU
LAVENDER BARN - School Lane. Tel: 0161 941 2153. Charming tea room and gift shop, closed Mondays. WA14 4TR

Linotype Silhouette

Shopping
Tiny village shop adjacent to the church for newspapers, basic provisions and local gossip. On the other hand, Little Heath Farm Shop (Tel: 0161 928 0520 - WA14 4SE) is a cornucopia of home-reared beef, lamb and pork, sausages, vegetables and New Cheshire potatoes. Dunham Massey Brewery (Tel: 0161 929 0663 - WA14 4PE) sell their own bottled beers (together with some wines and chutneys) from their premises on Oldfield Lane. On Station Road, Dunham Woodhouses, Dunham Massey Farm (sic) sells home-made ice cream, and they are open daily, albeit only in the afternoon - Tel: 0161 928 1230.

Things to Do
DUNHAM MASSEY - Tel: 0161 941 1025. House (including 17th century watermill) open late Feb-early Nov, Sat-Wed, 11am-5pm. Gardens, parkland, restaurant and NT shop open daily throughout the year. Admission charge. Dunham Massey Hall was bequeathed to the National Trust by the 10th Earl of Stamford in 1976. It features one of Britain's most sumptuous Edwardian interiors. Over thirty rooms are open to the public housing an admirable collection of furniture, paintings and Huguenot silver. Heaps of family history: the 7th Earl infamously married a bareback circus rider. Fallow deer roam in the 250 acres of delightful parkland. WA14 4SJ

Connections
BUSES - services 5 and 38 link Dunham Town with Altrincham in one direction and Lymm and Warrington in the other. 3 buses per hour Mon-Sat, hourly Sun. Handy bus stop outside main gate of Hall. Tel: 0871 200 2233.

Oldfield Brow   Map 11
Facilities for canallers include a convenience store, post office/pharmacy (with cash machine), sandwich bar (Tel: 0161 929 8100), Tandoori (Tel: 0161 924 0161) and fish & chips/Chinese (Tel: 0161 941 1127).

Connections
BUSES - services 13 & 14 run half-hourly (hourly Sun) to/from Altrincham.

12 BRIDGEWATER CANAL Sale 4.5mls/0lks/1.5hrs

THE Bridgewater Canal traverses the apparently boundless suburbs of Altrincham and Sale, an urban sprawl broken only by the green corridor of the Mersey's flood plain.

It is odd to reflect that none of these buildings were here when Brindley charted the Bridgewater cut across the flat peat mosses towards the Runcorn Gap in the middle of the 18th century. Indeed, this area remained predominantly agricultural (its fecundity enhanced by daily cargoes of 'night soil' from inner Manchester) until construction of the Manchester, South Junction & Altrincham Railway brought an explosion of house building in its wake from the 1840s onwards. Before they actually built the railway there were proposals to convert the canal into one. Thankfully, nothing came of that, but the trains did bring an end to the passenger packet boat services hitherto operated with aplomb.

Prior to the coming of the railway, Sale was just a barren waste with a rifle range. When Napoleon was threatening to invade in 1804, the Duke of Gloucester held a grand review of regulars and volunteers on the open expanse of Sale Moor, as vividly described by Mrs Linnaeus Banks in her 1876 novel *The Manchester Man*.* Similarly, Altrincham was a modest market town of less than four thousand souls. Fifty years later they had amassed a combined population of 100,000. The train service was intensive and electrified as early as 1931. Nowadays it is operated as part of Manchester's 'Metrolink', an expanding rapid transit system which, when the first routes to Altrincham and Bury opened in 1992, restored trams to the city's streets for the first time in 43 years.

Broadheath (aka Altrincham) Bridge* in its present guise dates from 1935, and exudes Art Deco overtones to prove it. In contrast, a few hundred yards short of it to the west, stands an old cotton warehouse of 1833. Dwarfed by Urban Splash's projecting apartment blocks at the rear of their redevelopment of the Budenberg Gauge Company's offices, it is immensely sad that such an intrinsically handsome building (note the bricked-in loading archway) should be abandoned and in such a bad state of repair, and it is difficult to understand why it could not have been incorporated into such an otherwise excellent regeneration scheme.

Remnants of two old railways (LNWR & CLC respectively) cross the canal before it swings round to run alongside the Manchester, South Junction & Altrincham Railway and its numerous Metrolink stops. Walton Park boasts a railway of quite a different kind, the miniature line of the Sale Area Model Engineering Society whose circular elevated track is designed to accept locomotives of both three and a half and five inch gauge. They get up steam most Sunday afternoons.

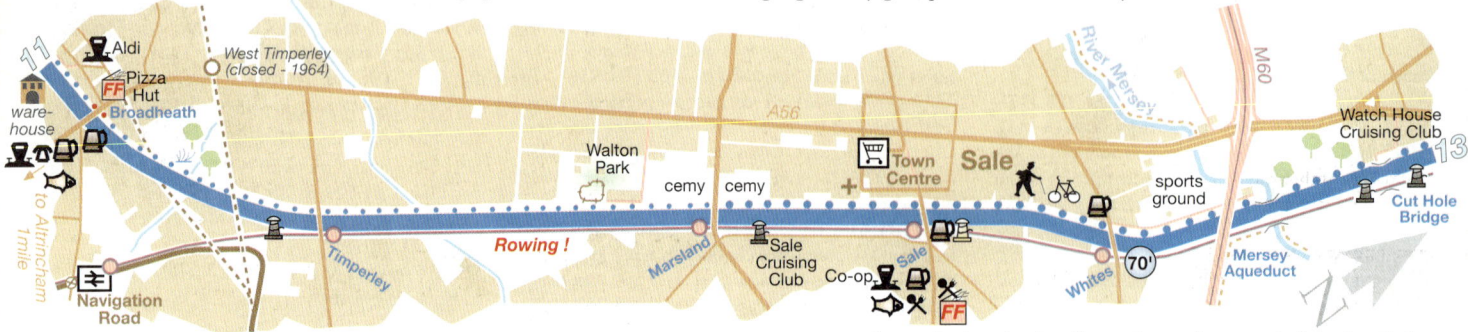

* Whose entertaining narrative also takes you to Whaley Bridge - Map 17B

*access is via a slender alley at the north-east end of the Sea Cadets

Trafford Rowing Club, established in 1957 operate out of premises which were once used as a canalside grain store. Two years after the club's formation, they faced competition for the Bridgewater Canal's turgid waters in the shape of Sale Cruising Club.

Sale Brooklands Cemetery was opened in 1862. Amongst its illustrious occupants are the scientist James Prescott Joule, and the husband of suffragette Emmeline (and three equally committed daughters), Richard Pankhurst. Curving round to pass beneath the M60, the canal passes Manchester University's athletics track and rowing club. Beyond the M60 [...] Acre Farm, premises of the Society for Abandoned Animals

who take in and try to rehome unwanted dogs, cats and rabbits. Sale Water Park has been created out of gravel extraction cavities associated with the construction of the motorway and is popular with windsurfers, water skiers and small boat sailors.

Watch House Cruising Club occupy a former length-man's cottage overlooking a series of arches which carry the canal over the Mersey Valley. Boat clubs are a feature of the Bridgewater Canal these days, but it wasn't until 1952 that pleasure cruising was permitted at all. In the Fifties the Watch House was base for a horse-drawn hotel boat operation which plied canals all over the country.

(Map 12)

[...]sed as a residential [...] Altrincham hasn't [...] independent [...] round the [...]ish list, [...]ardingly [...]d Brow,

WHEN this Canal Companion first [...] would urge travellers on the Trent & [...] eyes peeled for the passage of coasta[...] Weaver. Nowadays it would make m[...] to watch out of UFOs, for waterborne trade on t[...] out at the end of the 20th century. What is it in [...] what flaw in transportation theory, that leaves a [...] harmonious mode of transport, capable of sus[...] infrastructure like the Weaver, devoid of [...] Inland shipping apart, this is an extrem[...] & Mersey. Rolling farmland, interspers[...] characterise the canal's progress along a [...] valley. Picturesque to a fault, bu[...] summer of 2012 (together) a significant [...] maintenance) a significant [...] resulted in the newly for[...]

Indian [...]
PHANTHONG - [...]
Town centre Thai. WA14 1R[...]

Shopping
Altrincham can trace its status as a market town back to the 13th Century and remains a good shopping

centre as befits its prosperous suburbs. The handsome market hall dates from 1849 and operates on Tuesdays, Fridays and Saturdays for general merchandise, and on Thursdays for antiques. Abacus is a good secondhand bookshop on Regent Road.

Things to Do
TOURIST INFORMATION - Library, Stamford New Road. Tel: 0161 912 5931. WA14 1EJ

Connections
BUSES - frequent services (5, 37 & 38) to/from Warrington via Lymm. Tel: 0871 200 2233.
METROLINK - frequent trams to/from Manchester via Sale. Tel: 0161 205 2000.
TRAINS - hourly Northern service to Chester (via [N]orthwich) and Stockport/Manchester. Tel: 03457 [...]950.
[...]S - Trafford Cars. Tel: 0161 928 1111.

Map 12

[...] arguably best known for its rugby union club [...]nd share now with their rugby league [...], Salford City Reds, at the Salford City [...]ooking the canal, the cupola-topped, [...] Mary style Town Hall dates from the [...]and, unfortunately, sets an architectural [...] the town centre fails to sustain.

Eating & Drinking
BORELLO - School Road. Tel: 0161 962 4455. Italian just west of Sale Bridge. M33 7XY
J. P. JOULE - Northenden Road. (just east of Sale Bridge). Tel: 0161 928 9889. Wetherspoon. M33 3LF
KINGS RANSOM - Britannia Road (by Sale Bridge). Tel: 0161 969 6006. Canalside Greene King. M33 2AA
SOKRATES - Northenden Road. Tel: 0161 282 0050. Greek restaurant of Sale Bridge. M33 2DH
TABLE 10 - Northenden Road. Tel: 0161 282 2212. Intimate restaurant just east of Sale Bridge. M33 3BR

Shopping
There's a Co-op convenience store yards east of Sale Bridge (and a Barclays Bank with ATM too) but the town centre lies to the west of the canal and incorporates a pedestrianised shopping centre called 'The Square'. Tesco, Sainsbury's and Aldi have supermarkets either side of School Road, but we were drawn to a nice little fishmongers on Claremont Road boasting that the occupants of its slabs came 'fresh from Fleetwood'.

Connections
METROLINK - frequent trams to/from Manchester and Altrincham. Tel: 0161 205 2000.

13 BRIDGEWATER CANAL Waters Meeting 3.5mls/0lks/1hr

WATERS MEETING is a misleadingly arcadian appellation for the benighted junction of the Bridgewater Canal's Leigh and Manchester lines; though, in mitigation, it must have been of much more rural aspect in the 18th century before being engulfed by the world's first industrial estate. Nowadays it's a gloomy spot where gangs gather, though don't let that deter you, mostly they are too absorbed in their own nefarious dealings to be a threat to canal travellers.

Progressing in an easterly direction, the suburbs begin to make way for industry and vice versa. Longford Road Bridge marks the temporary terminus of the Bridgewater Canal from Worsley between 1761 and 1765, before Castlefield was reached. On the outside bend, south of the bridge, stood Rathbone's, a notable drydock and boatbuilding yard. When the Bridgewater Canal opened it stimulated the development of its hinterland for market gardening. It seems difficult to believe now that Stretford was once a centre for pig rearing.

East of Waters Meeting, the canal skirts a busy rail/road container depot. Logistics have come a long way. Via the Channel Tunnel, freight from the south of Spain can be unloaded here in the same time that it used to take to arrive from East Lancs.

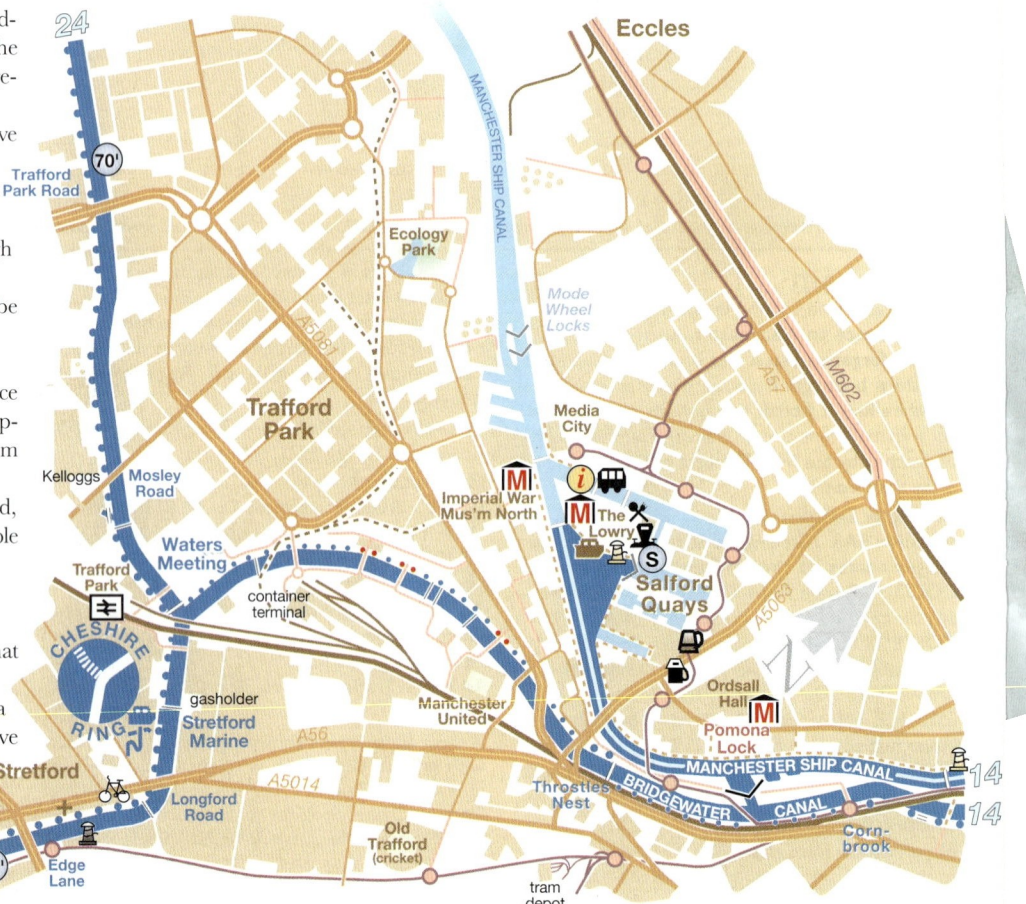

Manchester United's iconic Old Trafford stadium overlooks the canal. It must be an exhilarating experience to pass here when a game is taking place, with seventy thousand souls roaring on their team. On the towpath side mooring pins recall where barges, travelling down from collieries in the vicinity of Worsley, Leigh and Wigan, delivered coal to Trafford Park power station until 1972. The canal traffic's decline mirrored the football team's. United were at their post-war nadir, being relegated, at the end of the 73/74 season to the wilderness of Division Two. It's a shame that trade on the canal could not be regenerated the way the team was by (the canny Scot) Sir Alex Ferguson. Jose Mourinho (the cheeky Portuguese) was in charge at the time of our latest update.

The Liverpool Warehousing Company's variously dated cotton warehouses overlook the canal and remind us of lost trade as it threads its way under numerous concrete road crossings. At Throstles Nest Bridge the towpath changes sides. Extending the Metrolink tram network to Salford Quays and Eccles has had a remarkable effect on this once heavily industrialised area of Manchester. The trams snake across the canal on a bridge, but we're all far too late to see this environment at its most interesting. Half close your eyes and imagine Pomona's once busy dockland: swinging cranes, shouting stevedores, tank engines shunting wagons along tracks set in cobbles: ropes being thrown, oaths sworn, hooters blown. In the Fifties Manchester was the third busiest British port after London and Liverpool, twenty million tons of cargo being handled annually, and the docks at Salford and Manchester were awash with shipping from all over the world. Now the scrub covered wastelands that remain are awash with sadness in the psyches of those who wish they'd witnessed such scenes.

Pomona Lock was installed in 1995 as a replacement for Hulme Locks (Map 14). Given sufficient boating experience and confidence, arrangements can be made with the Bridgewater Canal Company (see page 101) to negotiate the lock and gain access to the Irwell. Downstream you can reach Salford Quays; upstream, Manchester Cathedral (Map 14) or (theoretically!) the Manchester Bolton & Bury Canal.

Waters Meeting - Barton upon Irwell

North-west of Waters Meeting the Bridgewater Canal is immediately submerged in the precincts of Trafford Park industrial estate which keeps it company all the way to Barton, though, it has to be said, the canal seems to eschew dialogue with the businesses on its banks, being brutally separated from them by tall, barbed wire-topped fences. In the majority of cases the canal didn't play much part in the development of these industries, but two notable exceptions were the carriage of chemicals by Cowburn & Cowpar narrowboats to and from Courtaulds, and the use of barges to bring grain from Salford Docks to Kelloggs who built a works here just prior to the Second World War as part of an initially ill-timed drive to enter the European health food market. The use of barges continued until 1974. Merchant ships carrying up to 10,000 tons of American grain would tranship their cargoes into a fleet of Bridgewater barges. A roving bridge carries the towpath over the arm into which the barges entered the works for unloading by suction into the high silos. A million cartons of Corn Flakes, Frosties and Coco Pops are turned out here per day, and it's heartening to learn that Kelloggs, after years of relying on lorry transport, are making increased use of containerised ingredients conveyed aboard ship (see also page 19) along the MSC to Irlam.

Prior to the Ship Canal's opening in 1894, Trafford Park was the sylvan country seat of the de Trafford family who had held sway hereabouts since the 11th century. Fierce opponents of the Manchester Ship Canal - on the grounds that its construction would render Trafford Hall uninhabitable - the family put the estate on the market in 1896. Amidst considerable controversy, it was purchased by the developer Ernest Hoole, who joined forces with Marshall Stevens, General Manager of the MSC, to create Trafford Park, the world's first industrial estate. At its zenith during the Second World War, Trafford Park employed over seventy thousand people. Trafford Park Village was developed to house the workforce. It was modelled on the American gridiron pattern, with streets and avenues given numbers rather than names. A gas-powered tramway provided transport facilities for both passengers and freight.

14 ROCHDALE CANAL Manchester 2mls/9lks/3hrs

SALFORD

MANCHESTER

N

Eating & Drinking
1 Akbar's
2 Albert's Shed
3 Briton's Protection
4 Cask
5 Croma
6 Dimitri's
7 Dukes 92
8 Jolly Angler
9 Knott Bar
10 Little Yang Sing
11 Manchester House
12 Peveril of the Peak
13 Sapporo Teppanyaki
14 TNQ
15 The Wharf

Limit of Navigation !

Cathedral
National Football Museum
Printworks
SHUDEHILL
Shambles Square
Northern Quarter
Royal Exchange
Arndale Market
Arndale Shopping Centre
St Ann's Square
OLDHAM STREET
HIGH ST
MARKET ST.
CROSS STREET
DEANSGATE
King Street
JOHN DALTON ST.
CHAPEL STREET
BLACKFRIARS ST.
Parsonage Gardens
BRIDGE STREET
Lowry Hotel
Tesco
Salford Central
IRWELL STREET
RIVER IRWELL
TRINITY
WATER STREET
WATER WAY
EAST ORDSALL LANE
Manchester Bolton & Bury Canal
Wilburn Street Basin
People's History Museum
Spinningfields
Marriott Hotel
Opera House
John Rylands Library
Albert Square
Town Hall
Central Library
Midland Hotel
Art Gallery
China-town
St Peter's Square
MOSLEY STREET
PRINCESS ST.
PORTLAND STREET
SACKVILLE
CHORLTON ST
MINSHULL
AYTOUN ST.
Piccadilly Gardens
Piccadilly
DALE ST.
DUCIE ST.
STORE STREET
LONDON ROAD
Piccadilly Basin
Ald
Crown Court
Gay Village
FF
CANAL ST.
GRANBY ROW
WHITWORTH STREET
GRANBY ROW
Palace Theatre
Palace Hotel
Bridgewater Hall
LOWER MOSLEY ST.
PETER STREET
QUAY STREET
GT. BRIDGEWATER ST.
Great Northern
Central Convention Complex
Beetham Tower
Castlefield Hotel
YMCA YHA
MoSI
Liverpool Road
Castlefield Roman Fort
Potato Wharf
former Hulme Locks
REGENT ROAD
Course of Manchester & Salford Junction Canal
BRIDGE WATER
CANAL
Castlefield Junction
C'field Quay
Deansgate
MANCUNIAN WAY A57M
River Medlock
Home
Oxford Road
OXFORD STREET
WHITWORTH STREET WEST
ROCHDALE CANAL
13

this map to approximate scale: 3½ inches to a mile
for details of facilities in Manchester turn to pages 31 and 32

28

O VER thirty years ago, when we published the first edition of the *Cheshire Ring Canal Companion*, received wisdom discouraged tarrying in Manchester. You moored overnight in Romiley or Sale, cast like a camel train furtively off at dawn, and did a twelve hour stint at the tiller. Manchester was so synonymous with decay and vandalism that it was regarded by the majority of canal travellers as a necessary evil; a temporary aberration in the otherwise predominantly scenic character of the Cheshire Ring. Times and perceptions have changed! Revitalisation and refurbishment - both on the canals and throughout much of the adjoining urban environment - have transformed the canal journey through Manchester into (in our opinion at any rate) one of the highlights of the Cheshire Ring. Personally, we find these canals so attractive and intriguing now that we would gladly moor up for a week, content to use the boat as a base for exploring the many fascinating aspects of this vibrant city.

The Canals of Castlefield

Doyen of the Canal Age, the Bridgewater arrived at Castlefield in 1765: George III was on the throne and America was still a British colony. Forty years were to pass before the Rochdale Canal came to join the Bridgewater. Everyone knows that the latter was dug to carry the Duke of Bridgewater's coals from his mines at Worsley (Map 24) to the market place of Manchester, but the canal soon developed as a general carrier and numerous warehouses sprang into being at Castlefield for the storage of multifarious cargoes. Several remain intact. Dominating the junction, the Merchants Warehouse has been refurbished, as has the vast Middle Warehouse overlooking the arm leading to the River Medlock - known as Castlefield Quay now and offering the city's best visitor moorings to our way of thinking. At the end of this arm, at a point where the original canal tunnelled beneath the sandstone outcrop, stands the Grocers Warehouse. Dating from the 18th century, the building originally consisted of five floors with a central arch over the canal. Later a second arch was added. An ingenious system of sluices fed a water wheel which drove the warehouse's lifting machinery. Two floors topped by a viewing promenade have been reconstructed, whilst the former bays are

fronted by a pair of not entirely appropriate nor aesthetically pleasing metal lift bridges.

But it would be churlish to be pedantically critical of Castlefield. Some observers have cited the Merchants Footbridge of 1995 (which spans the junction) as being out of keeping with the largely 19th century environment surrounding it, but we feel it compliments the adjoining railway structures without ingratiatingly replicating them. It is wider at its centre to provide space for pedestrians to pause to take in the view, and was inspired by Calatrava's Ripoll Bridge in Gerona. The adjoining railway viaducts are of enormous aesthetic appeal as well. See how the original masonry arches have been parodied in cast iron. One set of tracks carries conventional trains, the other, higher level, forms the Metrolink tramway approach to the former Central Station, reborn as a convention complex. A series of arms extend beneath these railway arches towards Liverpool Road and the Museum of Science & Industry. This area was known as Potato Wharf, a name reflecting the use of these arms as an unloading point for market garden produce brought in by boat from the farmlands of Cheshire.

The Rochdale Nine

Better maintained than in the past, though still considered hard work, the Rochdale Nine (locks 92-84) vouchsafes canal explorers with an off-beat, quirky and stimulating perspective of the city, an adventurous insight denied mere tourists. The bottom chamber is known as Duke's Lock because it was actually built by the Bridgewater company. Later it was renumbered in the Rochdale sequence. The 92nd lock on the canal's epic Trans-Pennine journey from Sowerby Bridge. Looming imperiously over all this is Manchester's tallest building, Beetham Tower, a mixture of the Hilton Hotel and private apartments. Those with a head for figures - if not necessarily heights - may care to know that it is five hundred and fifty-four feet high. Passing beneath yet another highly decorative railway bridge, the Rochdale Canal negotiates a short tunnel under Deansgate (variously known as Deansgate, Gaythorn or just plain Bridge 100) before

continued overleaf:

continued from page 29:

encountering, between locks 91 and 90, a split level boardwalk development of clubs and bars known as "Deansgate Locks" which occupies former railway arches. By Bridge 99 the Hacienda Apartments derive their name from the famous night club which stood on the site between 1984 and 1997. Heart of the 'Madchester' scene of that heady 'acid house' and 'rave' era, the club was closely associated with the band New Order, though it's worth noting that The Smiths played the club on several occasions and that Madonna's first UK appearance took place here in 1984.

Lock 89 marks the former junction with the Manchester & Salford Junction Canal, a section of which was rewatered to connect with the Bridgewater Hall, home of the celebrated Halle Orchestra. It seems to us that an opportunity was lost to provide city centre visitor moorings at the end of this arm where it opens out into a small basin beside the Bridgewater Hall. But a boom prevents entry and the basin has been cosmetically enhanced by a half-hearted fountain. As it passed beneath Deansgate, the M&SJC connected with the basement of the Great Northern Railway goods warehouse, now a retail and entertainment complex.

Manchester's very own Oxford Street spans the canal at Bridge 98, the view here being dominated by the terracotta tower of Sir Alfred Waterhouse's Refuge Assurance building, now the Palace Hotel. Bloom Street Power Station overlooks the pound between locks 88 and 87. It was built in 1902, primarily to provide power for Manchester's tramways, but surplus steam was harnessed to provide heating for shops and offices in the vicinity. It was even used to raise the curtain at the Palace Theatre.

Over Princess Street (Bridge 97) double-decker buses create a kaleidoscope of colour, enlivening the muted tones of massive textile warehouses refurbished as flats, hotels and restaurants. Leave the towpath and turn right into Whitworth Street and you'll be confronted by the astonishing Edwardian baroque of India House. Built in 1906, it featured in Adolphe Valette's painting of the same name.

Canal Street (along which towpath users must divert between bridges 97 and 94) lies at the heart of Manchester's Gay Village, venue of the annual Manchester Pride festival. Pavement cafes provoke an incongruous juxtaposition with a canal environment which was once the sole preserve of burly bargees who would scarcely believe their eyes at the scenes enacted here now. By Bridge 96, a Beacon of Hope sculpture is dedicated to all affected by HIV and AIDS. In neighbouring Sackville Gardens the sculptured figure of a man on a park bench holding an apple in his palm commemorates Alan Turing, codebreaker at Bletchley Park during World War Two, and father of computer science at Manchester University in the early Fifties. Turing ate an apple laced with cyanide in 1954 because his homosexuality had been exposed and he had become a victim of prejudice. It seems entirely apt, then, that prejudice has given way to pride.

Manchester's Crown Court overlooks the canal between Minshull and Aytoun streets. It dates from the 1870s and was designed in Venetian Gothic by Thomas Worthington who also did the Albert Memorial by the Town Hall. Over Bridge 93A many a doomed prisoner must have trudged to their fate.

The city's modern (and increasingly numerous) trams purr their way over Bridge 93, lending a continental aspect to the passing scene, before the canal briefly becomes subterranean. Lock 85 used to do a passable impression of what the Styx would have looked like if Brindley had ever been called in to make it navigable. It has, however, been illuminated in recent times. More, admittedly, as a deterrent to the 'rough trade' which this subfusc zone attracts, than to make life easier for boaters whose needs are invariably of secondary consideration. Oh, and don't worry if you do encounter shadowy loiterers, their thoughts will be elsewhere.

Emerging Orpheus-like from the underworld into the dank chamber of Dale Street Lock (No.84), boat crews may feel themselves entitled to a round of hugs and high-fives. Celebrations, however, will have to be curtailed if continuing eastwards, where more locks lie, devilishly in wait!

The Canals of Piccadilly

Once upon a time Piccadilly Basin (formerly Dale Street) boasted numerous wharves and warehouses. Now, in place of tethered barges, there are parked cars, and the past enjoys a joke at our expense. The basin adjoins the junction of the Ashton and Rochdale canals, a stone's throw from Piccadilly railway station on the south-eastern edge of the city centre. In amongst the tower blocks there are a few tangible remains of the Rochdale Canal's proud heritage: a crenellated entry arch off Dale Street itself; the canal company's offices; and a substantial millstone grit-built warehouse converted into offices. Its date-stone (1806) bears the initials 'WC', which at a guess relate to William Crosley, John Rennie's assistant. Round the side it plaintively reveals its water arches. On neighbouring Ducie Street the Place Apartment Hotel occupies a corresponding warehouse of the railway era. Piccadilly Basin offers good visitor moorings at the east end of the city centre. Canada geese are catered for too!

The Manchester, Bolton & Bury Canal

A short section of the Manchester, Bolton & Bury Canal was re-opened in 2008 as part of a redevelopment scheme in Salford. Access for boaters is via Pomona Lock (Map 13) and the River Irwell, so both the Bridgewater Canal Company and the Canal & River Trust will need to be contacted in advance to make arrangements. The MB&B was opened to broadbeam dimensions in 1797, though not at first linked to the Irwell in Salford. The main line to Bolton measured eleven miles, the branch to Bury from Prestolee was four and three-quarter miles long. There had been consideration of linking the canal to the Rochdale Canal by way of an aqueduct over the Irwell before a flight of six locks down into the river were constructed in 1808. Two centuries later the link was re-established with some panache; though, on the occasion of our most recent reconnaissance, the MB&B lay silted, weed-bound and moribund, an initiative melancholically ahead of its time.

Manchester   Map 14

The pace of change in Manchester is so furious that each new edition of this guide feels like a fresh introduction to a relative stranger. Manchester seems like a city on an ever upward trajectory, reinventing itself with a fervour that one can only admire, even if it borders on the obsessive. The sadness, from the canaller's viewpoint, is that Manchester's act of embrace with its waterways appears not to be returned with equal affection. Unlike Birmingham, say, canalside redevelopments here appear less well integrated; more a matter of passive reflection than passionate response. The best part of twenty years have passed since Castlefield's canals were refurbished, though boat movements remain at such a relatively low level that it sometimes feels like a party which disappointingly few invitees have bothered to attend.

Architecturally, however, Manchester remains

one of the most handsome cities in the post-industrial world, and we would urge you to moor and explore; nothing's very far away, the city centre being astonishingly compact and approachable. 'Musts' include the three squares: Albert, with Waterhouse's imposing Town Hall; St Ann's, an oasis of relative peace in the city centre; and St Peter's, dominated by the circular Central Library and flamboyant terracotta of the Midland Hotel, meeting place (4th May, 1904) of a certain Mr Rolls with a certain Mr Royce. More esoterically, try venturing into the rag trade zone east of Piccadilly and thence to High Street where the skeleton of the old Wholesale Fish Market features entertaining friezes illustrating fishing, landing and selling scenes. Manchester boasts thousands of buildings like this, evoking its plutocratic zenith. Thence wander down to the banks of the Irwell where, in the shadow of the modest cathedral lies Chethams Music School, Victoria Station and the stolidly confident Co-operative buildings. These are the parts of the city reassuringly Northern in character. The smoke palls may have been blown away by the winds of change, but there are moments when it is still possible to feel that you've walked on to a canvas by Adolphe Valette or his pupil L. S. Lowry.

continued overleaf:

continued from page 31:

Eating & Drinking

AKBAR'S - Liverpool Road. Tel: 0161 834 7222. Indian restaurant opposite MoSI. M3 4NQ
ALBERT'S SHED - Castlefield. Tel: 0161 839 9818. Classy restaurant housed in old Bridgewater Canal Co. warehouse. Open from noon weekdays, 10.30am weekends. Access Lock 92/Bridge 101. M3 4LZ
BRITON'S PROTECTION - Great Bridgewater Street. Tel: 0161 236 5895. *Good Beer Guide* perennial and listed in the CAMRA National Inventory of Historic Pub Interiors. M1 5LE
CASK - Liverpool Road. Tel: 0161 819 2527. *GBG* listed pub for beer lovers. No food. M3 4NQ
CROMA - Albert Square (Clarence Street). Tel: 0161 237 9799. Italian restaurant open from noon. M2 4DE
DIMITRI'S - Campfield Arcade, Castlefield. Tel: 0161 839 3319. Greek tapas bar and taverna. M3 4FN
DUKES 92 - Castlefield Junction. Tel: 0161 839 8642. Longstanding canalside bar and restaurant renovated in 2016. Open daily from 11am, food served from noon throughout. Meat, fish & cheese sharing boards, sandwiches, small plates, mains, grills, pizza etc. Access Lock 92/Bridge 101. M3 4LZ
JOLLY ANGLER - Ducie Street. Tel: 0161 236 5307. Friendly basic boozer featuring locally brewed Hydes. Access from Ashton Canal Bridge 2. M1 2JW
KNOTT BAR - Deansgate. Tel: 0161 839 9229. *Good Beer Guide* listed bar built into the railway arches by Deansgate Tunnel easily accessed from Castlefield moorings. Wide range of beers. Food. M3 4LY
LITTLE YANG SING - George Street. Tel: 0161 228 7722. Renowned restaurant in 'Chinatown'. M1 4HE
MANCHESTER HOUSE - Bridge Street. Tel: 0161 835 2557. One of the city's most critically acclaimed restaurants. Lunches & dinners Tue-Sat. M3 3BZ
PEVERIL OF THE PEAK - Gt Bridgewater Street. Tel: 0161 236 6364. Famous green-tiled pub named after a stagecoach. Prized for its old-fashioned football table. Access Lock 89. M1 5JQ
SAPPORO TEPPANYAKI - Liverpool Rd. Tel: 0161 831 9888. Japanese restaurant opp. MoSI. M3 4JN
TNQ - High Street. Tel: 0161 832 7115. Acronym for 'The Northern Quarter', this stylish restaurant - located opposite the former fish market illustrated on the previous page - opens from noon daily, though note early close on Suns at 7pm. M4 1HQ
THE WHARF - Slate Wharf, Castlefield. Tel: 0161 220 2960. Brunning & Price new-build pub in warehouse vernacular handily placed for Castlefield visitor moorings. Open daily 11am, food from noon. B&P never fail to impress on our canal travels, and this couldn't be closer to the best moorings. M15 4SW

Shopping

The principal shopping area lies to the north of the Rochdale Canal. Lots of outdoor gear and cycle shops towards the Castlefield end of Deansgate. There's a handy Sainsbury's Express on Oxford Street by Bridge 98 and Aldi by Bridge 90 on the Rochdale Canal.

Things to Do

VISITOR INFORMATION CENTRE - Piccadilly Gardens. Tel: 0871 222 8223. M1 4RG
MUSEUM OF SCIENCE & INDUSTRY - Liverpool Road. Open daily 10am to 5pm. Tel: 0161 832 2244. MoSI is a first class celebration of Manchester's industrial prowess and scientific endeavour. Includes the first purpose-built passenger railway station in the world, opened in 1830 for the Liverpool & Manchester Railway. It's disappointing, however, that they have never spilled out across Liverpool Road to display some appropriately Mancunian canal craft. Free admission (donations welcome). Nice Warehouse cafe 8am (9am weekends)-5pm or alternative first floor bistro 11am-4pm. Shop. M3 4FP

CITY ART GALLERY - Mosley Street. Tel: 0161 235 8888. Access is free (though donations welcome) to this inspiring gallery, particularly strong on British painting. Shop and cafe. 10am-5pm daily. M21 3JL
HOME - Tony Wilson Place (off Whitworth Street West). Tel: 0161 200 1500. Film, theatre and visual arts centre. Cafe bar and bookshop. M15 4FN
PEOPLE'S HISTORY MUSEUM - Spinningfields. Tel: 0161 838 9190. A look at the lives of ordinary people in Britain over the last two hundred years. Much Trade Union material. M3 3ER
SIGHTSEEING BUS TOURS - 2hr open top bus tours taking in city and Salford Quays. Start from Chorlton Street on selected dates. Tel: 0800 288 8746.
NATIONAL FOOTBALL MUSEUM - Urbis Building, Cathedral Gardens. Tel: 0161 605 8200. M4 3BG

Connections

BUSES - Tel: 0871 200 2233.
TRAMS - Tel: 0161 205 2000.
TRAINS - Tel: 03457 484950.
TAXIS - Tel: 0161 228 7878.

Salford Quays   Map 13

Given a timely boost by the establishment of MediaCity, Salford Quays are a 21st century manifestation of Salford Docks: commerce, leisure and retail where there used to be ships and dockers.
THE LOWRY - Salford Quays. Tel: 0843 208 6000. Multi-art venue. What would the curmudgeonly old bachelor of Mottram have made of it? M50 3AZ
IMPERIAL WAR MUSEUM NORTH - Trafford Wharf Road. Tel: 0161 836 4000. 'How people's lives are shaped by war'... as if we didn't know! M17 1TZ
ORDSALL HALL - Ordsall Lane. Tel: 0161 872 0251. Remarkable Grade I listed Tudor hall. Admission free. Closed Fri & Sat. Small shop and cafe. M5 3AN

THE macho Ashton we interpreted in earlier editons of this guide has progressively given way to an almost effete canal. De-industrialised and drained of its former atmosphere, it has the feel now of a written confession extracted under coercion. Which is not to say it is not worth exploring. Canal travellers - especially those armed with Pearsons - are adept at prising the maximum pleasure from the minimum circumstance and can approach the Ashton Canal with their eyes wide open: not expecting to be attacked by feral gangs - as was sometimes the case in the not too distant past - but neither under the misapprehension that the perverse rewards of an assault course no longer await them.

But to be accurate (and when are we anything but?) it isn't so much the canal which has changed as its hinterland. The regeneration of east Manchester, kick-started by the Commonwealth Games of 2002, continues apace, if somewhat dampened by the multi-dip recession of recent years. There being too many wastegrounds, as yet

unfilled, and too many untenanted apartments, office suites and 'restaurant opportunities' to fully comply with the planners' and developers' undoubtedly progressive visions. Hopefully extension of Metrolink, which rubs shoulders with the canal at a number of points, to Ashton will prove as economically beneficial to the area as the opening of the Ashton Canal at the end of the 18th century.

Connection with the Rochdale Canal at Ducie Street was made in 1800, and it is from these same (if much altered in ambience) environs that those exploring the Ashton depart (or, of course, arrive) today, dog-legging (Map 14) through bridges 1 and 2; the latter carrying vertiginously cobbled Jutland Street which L. S. Lowry sketched in 1929 when it was known as Junction Street. Paradise Wharf precedes Store Street Aqueduct (another subject matter of Lowry's) dating from 1798 and quite possibly the first such structure to be built at a 'skew' angle. Originally it spanned a watercourse known as Shooters Brook, but as the area developed industrially the brook was culverted and a road built over it. Piccadilly Village is a rather tepid and inscrutable approach to canalside redevelopment, token cranes and 'cast-iron' footbridges hardly constituting vernacular veracity. If urban canals such as the Ashton have a role to play in the 21st century then they should take their cue from Will Alsop's 'Chips' apartment block erected by

continued on page 36

Key

1 site of Robertsons Jam Factory
2 former packet boat house

⚠ anti-vandal 'handcuff' keys required for most of the paddle gear and swing-bridge 15

🐦 = New Islington Marina

Castlefield 1

Beetham Tower

Alan Turing

Jutland Street

Hulme

Castlefield 2

Deansgate

Manchester Interlude

34

Chips Ahoy

Guide Bridge 1

Fairfield Junction

The Strawberry Duck

Portland Basin

1855

BRIDGE VIEW CAFE

Aspects of the Ashton

Guide Bridge 2

35

continued from page 33:

Urban Splash alongside the short pound between locks 2 and 3. Not without its critics, and not entirely helpfully described in marketing-speak as 'funky, pink, brown and peach', one nevertheless feels that the builders of old Ancoats would have applauded and seen in this, and adjoining developments, an echo of their own architectural confidence. The Ashton Canal flung off myriad arms and basins to serve this once heavily industrialised part of Manchester. You can see on Pollard Street (by New Islington's Metrolink stop) the remains of a cast-iron overbridge. A welcome development is New Islington Marina, accessed through the electrically operated lift-bridge (CRT facilities key required) by Chips, and offering valuable visitor moorings. The canal needs such investments, because, truth be told, it looks shabby in comparison with the brave new world gradually growing around it. Better care needs to be taken of the locks and their associated machinery, and with the towpath as well, if the synergies between the canal and its environment are to march forward confidently and compassionately into the future together.

The Ashton Canal's notoriously arduous eighteen locks - for the most part fitted with hydraulic paddle gear and time-consuming anti-vandal locks - comprise the following flights: Ancoats (3); Ashton (formerly Beswick) (4); Clayton (9) and Fairfield (2). On aggregate they raise the canal by a hundred and sixty-six and a half feet, and, oddly enough, boaters find their spirits correspondingly rising the nearer they get to the top.

Fragments of the Ancoats so vividly described by Howard Spring in *Fame is the Spur* continue to manifest themselves; here a mill, there a backstreet, together with a corner 'chippy' cocking an etymological snook at Chips. You can try and take the chip shop out of Manchester but you can't take Mancunians out of their chip shops. Thus they still emerge from their apartment eyries, hair in curlers, to queue for a fish supper.

Overlooked by a gas-holder, the canal crosses the River Medlock, passes beneath the former Lancashire & Yorkshire Railway's connecting line between Ardwick and Miles Platting, and threads its way through 'Sportcity'. 19th century boatmen would be astonished by such scene changes. Where now cluster tennis, squash and athletics facilities, the Etihad Stadium of Manchester City Football Club and the National Cycling Centre's velodrome, previously stood textile mills, coal mines, chemical plants and an electricity generating station.

Between locks 10 and 11, 'Stockport Junction' marked the egress point of a five mile branch which never actually reached the centre of Stockport, petering out instead by the flour-caked, redbrick mills of South Reddish. Trade had petered out too by the 1930s though it wasn't officially abandoned until 1961. Infilled long ago, its course now forms part of National Cycle Route 60 and can be followed down through Openshaw and beyond across an aqueduct still straddling the old Great Central Railway's main line at Gorton.

Fairfield Locks (17 & 18) were duplicated in the heyday of the canal. The short pound between them is spanned by a stone footbridge (known locally as the camel's hump) with a strong family resemblance to the roving bridge at Portland Basin (Map 16). Also of interest is a boathouse which housed a packet boat prior to the coming of the railways. Near at hand is Fairfield Moravian Settlement, a secluded and strangely unworldly estate of Georgian town housing built about cobbled avenues lit by cast-iron lamps. The Moravians are a Protestant sect founded in Europe during the 15th century.

From Fairfield Junction a branch canal was built to Hollinwood on the outskirts of Oldham where its terminus lay less than a mile from the Rochdale Canal to which a link was mooted but never dug. Whilst collieries provided the branch with the bulk of its trade, a recreational element flourished at 'Crime Lake', a Tixall-like widening which attracted boathouses, tea rooms and fairground booths to its banks. The winter of 1854 was so bad that the lake froze over for three months and stalls were erected on the ice to cater for skaters. A property development has seen the first few hundred yards of the Hollinwood Branch rewatered and provided with a mooring basin. We will miss the fruity aromas emanating from Robertsons' jam factory by Bridge 18, closed since the last edition. The company's golliwog logo had proved an embarrassment in later years, and Premier Foods its owners discontinued the brand in favour of Hartleys.

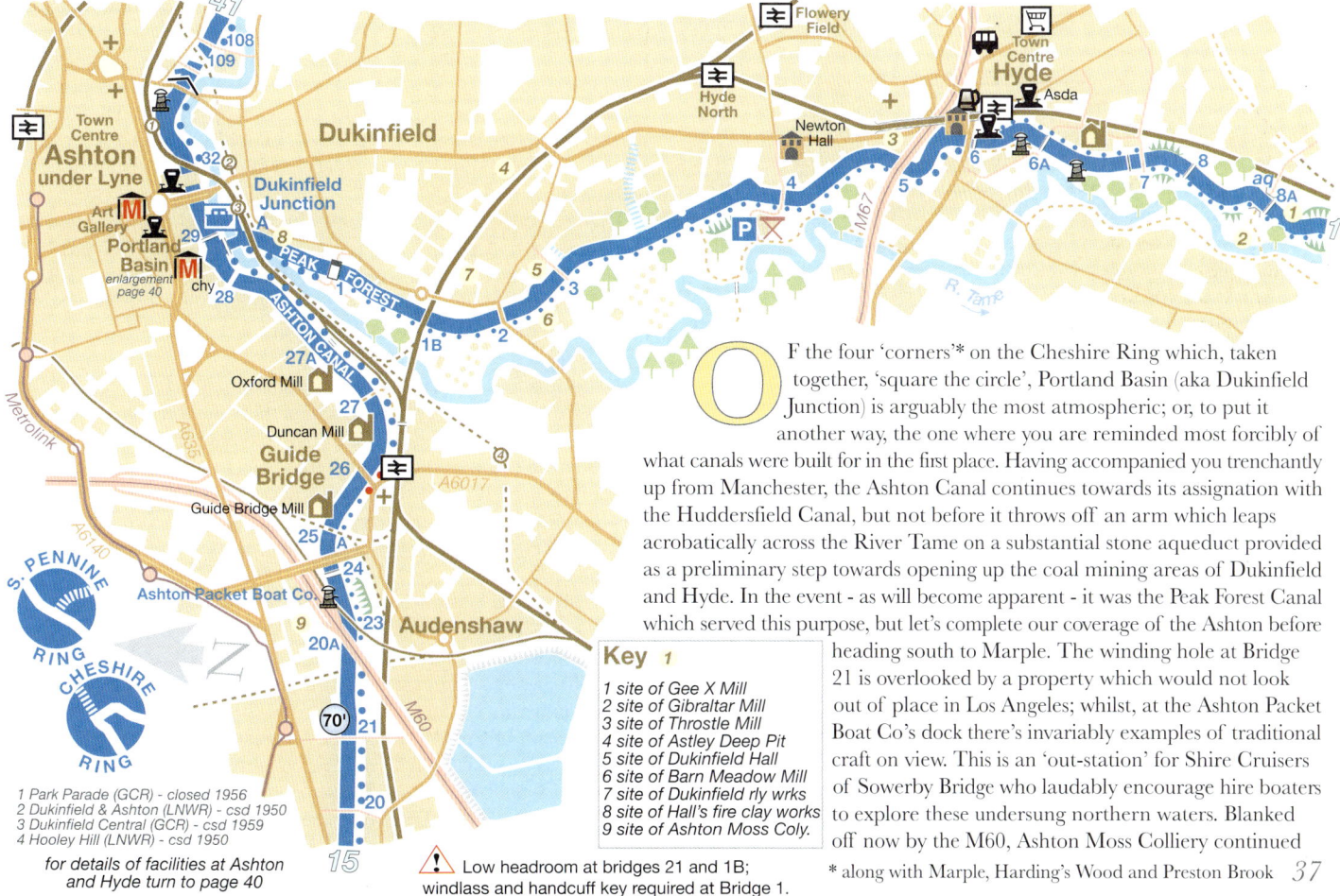

Flowery Field

Town Centre Hyde

Asda

Hyde North

Newton Hall

Town Centre Ashton under Lyne

Art Gallery

Portland Basin
enlargement page 40

Dukinfield Junction

Dukinfield

DUKINFIELD

PEAK FOREST

ASHTON CANAL

Oxford Mill

Duncan Mill

Guide Bridge

Guide Bridge Mill

Ashton Packet Boat Co.

Audenshaw

S. PENNINE RING

CHESHIRE RING

Metrolink

R. Tame

Key 1

1 site of Gee X Mill
2 site of Gibraltar Mill
3 site of Throstle Mill
4 site of Astley Deep Pit
5 site of Dukinfield Hall
6 site of Barn Meadow Mill
7 site of Dukinfield rly wrks
8 site of Hall's fire clay works
9 site of Ashton Moss Coly.

1 Park Parade (GCR) - closed 1956
2 Dukinfield & Ashton (LNWR) - csd 1950
3 Dukinfield Central (GCR) - csd 1959
4 Hooley Hill (LNWR) - csd 1950

for details of facilities at Ashton and Hyde turn to page 40

⚠ Low headroom at bridges 21 and 1B; windlass and handcuff key required at Bridge 1.

O F the four 'corners'* on the Cheshire Ring which, taken together, 'square the circle', Portland Basin (aka Dukinfield Junction) is arguably the most atmospheric; or, to put it another way, the one where you are reminded most forcibly of what canals were built for in the first place. Having accompanied you trenchantly up from Manchester, the Ashton Canal continues towards its assignation with the Huddersfield Canal, but not before it throws off an arm which leaps acrobatically across the River Tame on a substantial stone aqueduct provided as a preliminary step towards opening up the coal mining areas of Dukinfield and Hyde. In the event - as will become apparent - it was the Peak Forest Canal which served this purpose, but let's complete our coverage of the Ashton before heading south to Marple. The winding hole at Bridge 21 is overlooked by a property which would not look out of place in Los Angeles; whilst, at the Ashton Packet Boat Co's dock there's invariably examples of traditional craft on view. This is an 'out-station' for Shire Cruisers of Sowerby Bridge who laudably encourage hire boaters to explore these undersung northern waters. Blanked off now by the M60, Ashton Moss Colliery continued

* along with Marple, Harding's Wood and Preston Brook

in use until 1959.

The eastern end of this canal is characterised by knotted railways and textile mills both past and present. Railways were attracted to Guide Bridge like iron filings, notably the illustrious Great Central, whose renowned route between Manchester and Sheffield through Woodhead Tunnel passed this way. Had Sir Edward Watkin, its ambitious chairman, had *his* way, this would have formed a through route to Paris, a century before the Channel Tunnel was built. In the early 1950s it became the first main line in the country to be electrified. But to many people's lasting regret, this pioneering route was closed to passengers in 1970, and to freight just over a decade later. When we were researching the first edition of this guide back in 1981, disconsolate rows of electric locomotives were lined up in the sidings at Guide Bridge awaiting the breaker's torch. Much rationalised now, Guide Bridge's current claim to railway fame is that it is a stopping point for the Friday morning only, uni-directional, Stockport to Stalybridge service, retained to save the railway operators from going through the tedious and contentious procedure of applying legally to close the line.

Bridge 27A carries the Ashton Canal's much puddled towpath over what was once a short arm called Princess Dock, from which a packetboat operated. Bridge 28 is overlooked by the 210ft high Junction Mills chimney of 1867. Octagonal, and unusual for its 'tulip' shaped crown, the chimney was purchased for a pound by Tameside Council to prevent its demolition along with the mill it served, and now provides an appropriately plangent landmark for Portland Basin.

The canalscape at Portland Basin is immensely appealing. Everywhere you look something catches the eye. Pride of place goes to an extensive, three-storey warehouse originally built by the Ashton Canal Co. in 1834, destroyed by fire in 1972, but rebuilt to house a museum devoted to Tameside's rich industrial and social history. Directly opposite lies the aforementioned arm, handsomely spanned by a slender, stone-built roving bridge dated 1835. And then, once across the stone aqueduct which spans the Tame, you are officially on the Peak Forest Canal, and anticipating a whole new and exciting chapter of canal exploration.

Peak Forest Canal

Astonishing, to recall, that fifty years ago all three canals converging on Ashton-under-Lyne were officially abandoned. It took a crusade by canal enthusiasts to both lobby authority to revive these waterways, and to provide the muscle at the mucky end of their proposals. The Ashton and Peak Forest canals were reopened to navigation in 1974 (the Huddersfield had to wait another twenty-seven years) but only after tens of thousands of voluntary man hours had been devoted to removing detritus from the beds of both canals.

Fourteen and a bit miles long, and with sixteen locks concentrated into one flight, the Peak Forest Canal was constructed between 1794 and 1803, the final five, drawn out years concerning the locks at Marple being delayed by financial constraints. Its most important and profitable traffic flow was in limestone from Derbyshire quarries, but coal and cotton were also significant cargoes.

For its first four miles the Peak Forest finds itself journeying along a shoulder above the River Tame which, in those innocent days before the Heath Government's invention of a spurious county unimaginatively named Greater Manchester, formed the ancient boundary between Lancashire and Cheshire. This Tame - not to be confused with the one which travellers on the Coventry Canal encounter - rises on the moors above Saddleworth and merges with the Goyt to form the Mersey at Stockport.

Initially it feels as if the canal will have its work cut out to extricate itself from the canker of urbanisation, but canals - as you well know - have a facility for shedding their skin, and this one soon gets merrily into its stride, offering hints, here and there, of past glories. Good old BW's last bridge numbering initiative has led to a peculiar sequence as the canal passes beneath bridge 'A', '1' and '1B' in quick succession. But we must take them as we find them, for to cling on to the original, more logical progression would only lead to further confusion.

Revitalised by the Portland Basin Marina, the old Alma Street (or Garforth's Private) Branch extended eastwards immediately beyond the aqueduct. Immediately afterwards the canal is crossed by an ornate railway bridge dated 1845. Dukinfield Central station's platforms extended over this bridge until their demolition - nine years after the station's closure - in 1968. Scarcely any trace remains of the rival London & North Western Railway's line which crossed the canal on a parallel girder bridge. Gone too is Hall's fire clay works which closed down in 1967. Manufacturers of chimney pots, sanitary pipes and terracotta ridge tiles, they used boats in the heyday of the canal to ferry in raw clay from Bollington.

Bridge 1 is a lift bridge (requiring an anti-vandal key and a windlass to operate) which carries a track down over the canal to Plantation Farm which was on the market when we last passed; all four bedrooms and four acres of it. A nearby plaque recalls that Mary Moffat was born on the farm in 1795, that she became an African missionary, and that she was* an inspiration to her son-in-law, none other than David Livingstone.

Irrationally renumbered 1B, the bridge which carries the old Great Central main line from Manchester to Sheffield over the canal offers scant headroom to boaters. It is interesting to note, however, that the catenary masts and cross girders hark back to the original 1500 volts DC scheme of the 1950s.

With sewage farms on one side, and anonymous industrial units on the other, the canal proceeds southwards towards Hyde. The National Trust were approached after the Second World War to see if they would consider taking Dukinfield Hall under their wing, but they seemingly lacked sufficient enthusiasm. The Duckenfield (sic) family's most famous son was Lieutenant Colonel Robert Duckenfield who was appointed Commander of Cromwell's army in the north-west in 1648. There is a memorial to him in St Lawrence's church, Denton. By Bridge 3 stood a basin serving coal mines at Dewsnap and Astley, linked to the canal by tramway. Astley Deep Mine lived up to its name and was notorious for its safety record. In 1874 fifty four men and boys lost their lives in a huge explosion.

Five minutes on foot from Bridge 4 stands Newton Hall, a reconstructed half-timbered house which dates back to the 14th century. In the opposite direction lie the intriguingly named Jet Amber Fields, a public open

*somewhat rarely!

space frequented by kite flyers and other dreamers.

When the canal had to be realigned to accommodate the M67 motorway, thankfully Hyde Wharf with its loading shed and handsome warehouse was spared; the latter now finding unlikely use by the NHS. The towpath changes sides between bridges 6 and 7 to preserve the privacy of one of the promoters of the canal who dwelt at Hyde Hall. This length is overlooked by Joseph Adamson's engineering works, manufacturers, amongst other items, of many a textile mill boiler. Witness the judicious use of space by the cantilevering out of the works almost to the towpath's inner edge.

The Tame's inherent power proved a boon to 18th century millowners before steam power took over. Sometimes the location of industry is an accident of history, more often than not there are cogent reasons for specific areas being closely associated with a particular activity. The manufacture of cotton goods established itself on this side of the Pennines because the climate was one of high humidity; because the local water-courses provided both power for the looms and soft water for washing and dyeing; because coal was at hand with the advent of steam; because there was sufficient population to provide the workforce; and finally because the proximity of the Mersey ports gave access to world markets. Like a pair of worn out bloomers, the bottom fell out of the cotton industry as man-made fibres became more fashionable and less expensive to produce. Furthermore, it became apparent that all but the top strata of British society could dress itself more efficiently in garments manufactured in the sweat shops of the Far East. So the mills were made redundant. Laid up like obsolete oil tankers on a Cornish tidal creek. The maritime analogy is apt. Like ships the mills bore proud names. They lay, chimneys belching smoke, in the folds of the landscape like ocean going vessels in a swell. Whilst in their galleries hundreds of human beings once toiled 'below deck' to earn a meagre living. Two vanished examples of the breed lay adjacent to Apethorn Aqueduct by Bridge 8A. The neighbouring village of Gee Cross was an outpost of Manchester's trolleybus system, all of eight miles from Piccadilly. The boxer Ricky Hatton lives there now.

Ashton-under-Lyne   Maps 16/41

It's perhaps not widely remembered that Ashton-under-Lyne was name-checked in Flanders & Swann's *The Gnu Song* of 1960. But since when did anything appear in the Canal Companions that was widely remembered anyway? Once of Lancashire, now the foremost town of Greater Manchester's Metropolitan Borough of Tameside, Ashton's economy was built on textiles and coal-mining. Nowadays it revolves around retail and local government. The Town Hall exudes Corinthian-columned dignity. The parish church of St Michael was taken in hand by the Victorians but retains some notable 15th century stained glass devoted to St Helena. The trolleybuses (alas) left in 1966 but the trams (hurrah) are back. And last not least, this was the birthplace, in 1892, of the peerless travel writer H.V. Morton.

Eating & Drinking
BRIDGE VIEW CAFE - cafe/restaurant at Portland Basin Museum. Tel: 0161 343 6785. OL7 0QA
IL CAFFE ROMA - Market Avenue. Tel: 0161 258 6607. Cosy Italian in the old part of town. OL6 6BT

Shopping
The Market Hall is splendid, and it is quite possible to lose your womenfolk in it for hours on end while you attend to more serious matters ... like a meat pie from S. Williams & Sons. The Ladysmith shopping centre houses a cornucopia of well-known retailers, and is so-named, we imagine, because it's a 'relief' to leave it. An Asda supermarket spans the canal to the east of Portland Basin, whilst a branch of Lidl looms nearby, but you'd be mad to miss the market.

Things to Do
PORTLAND BASIN INDUSTRIAL MUSEUM - Portland Place. Tel: 0161 343 2878. Open Tue-Sun 10am-4pm, admission free. Provides a hugely

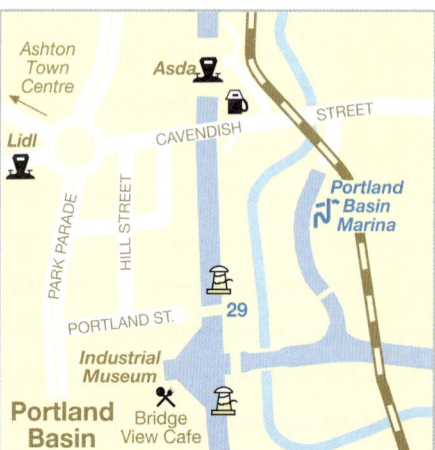

enjoyable insight into the area's industrial and social past. It is housed in a canalside warehouse built by the Ashton Canal Company in 1834. OL7 0QA
CENTRAL ART GALLERY - Old Street. Tel: 0161 342 2650. Housed in the Heginbottom Technical School and Free Library, the chief delight of this gallery is its collection of paintings by Harry Rutherford, a rather forgotten figure, though famous in the early days of television as a sort of forerunner to Rolf Harris. Pupil of Adolphe Valette (pages 30-31) and Walter Sickert, and contemporary of Lowry, one of Rutherford's best known paintings is *Mill Girls* which has for its backdrop the canal at Old Wharf and numerous mills. OL6 7SG

Connections
TRAINS - Northern services to/from Manchester Victoria, Stalybridge and Huddersfield. Tel: 03457 484950.
TRAMS - Metrolink. Tel: 0161 205 2000.
TAXIS - Radio Cars. Tel: 0161 330 2090.

Hyde   Map 16

Matters were simpler - not to mention prouder - when Hyde was a market town in Cheshire. You have Edward Heath and his ministers to thank for its subsumption into the ghastly amorphousness of Tameside in 1974. Other grubby associations with murderers (Shipman, Brady, Hindley et al) notwithstanding, Hyde is the sort of town you ought to stop and see, if only to witness what passes for life when the parade's passed by. Redemption comes with minor, joyful details such as the Let's Talk Tripe shop in the Market Hall. Hyde market was the subject of local artist Harry Rutherford's evocative painting *Northern Saturday*. A mural by the artist, which visitors are welcome to view, adorns the foyer of the substantial Town Hall.

Eating & Drinking
CHESHIRE RING - Manchester Road (adjacent Bridge 6). Tel: 0791 705 5629. *Good Beer Guide* listed local overlooking the canal which obviously inspired the pub's contemporary name. Six real ales! SK14 2BJ
PAGLIACCI - Market Street. Tel: 0161 351 1457. Homely Italian. Lunch & dinner daily ex Mon. SK14 1AH. *Fish & chips from Bosuns on Clarendon St.*

Shopping
Aldi and Asda within easy reach of Bridge 6. Open market and market hall daily (ex Sun)

Connections
BUSES - Hyde bus station lies on the way into town from the canal, but you will wait in vain for a number 210 trolleybus to Manchester - it last ran in 1963. Nowadays motor bus 330 runs usefully to/from Stockport and Ashton-under-Lyne. Tel: 0871 200 2233.
TRAINS - Northern run an hourly service (Mon-Sat - no Sunday trains) to/from Manchester and Rose Hill Marple, providing useful towpather links with Woodley, Romiley and Guide Bridge. Tel: 03457 484950.

FROM its post-industrial origins in the vicinity of Ashton-under-Lyne, the Peak Forest Canal gradually transmogrifies itself into one of the loveliest in the country, and by the time it skirts Romiley the reinvention is well under way. Two tunnels heighten the sense of expectation and adventure. Woodley Tunnel is prefaced (from the north) by an imposing stone-built, skew-arched, ivy-clad and pigeon-haunted railway bridge erected by the grandiloquently titled Cheshire Lines Committee, whose tracks Cheshire Ring travellers encounter on a number of occasions. Watch out for the containerised rubbish trains which rumble lugubriously across this bridge, conveying Manchester's refuse to a big hole on the outskirts of Scunthorpe. Talking of rubbish, CRT's signboarded instructions that Woodley Tunnel (aka Butterhouse Green) is the subject of 'two-way working' is surely contradictory. And, by the way, it gets a bit wet in the middle!

Bridge 13A carries another railway line; though in fact it once carried two, the once useful link with Stockport Tiviot Dale having succumbed to Marples & Beeching in 1967. More like a short tunnel, the structure's echoing properties are well worth indulging in. Bridge 14 provides access to Romiley. The Boat House Inn once abutted the north-west corner of the bridge, an important stop on a fly-boat passenger service which

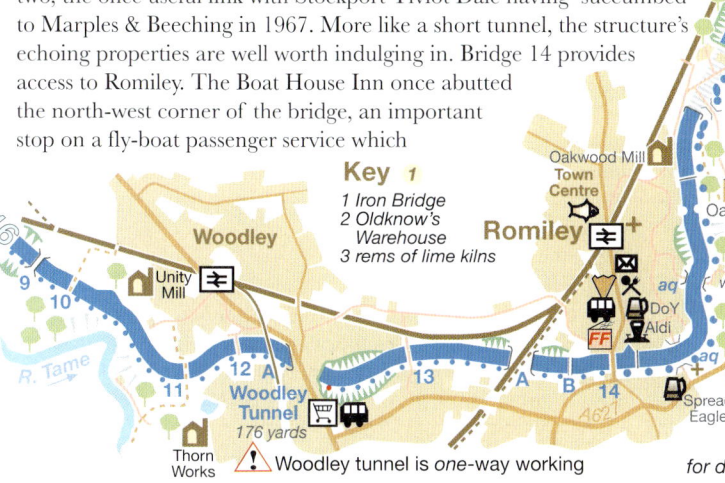

Key 1
1 Iron Bridge
2 Oldknow's Warehouse
3 rems of lime kilns

Woodley tunnel is *one-way* working

linked Dukinfield with Marple in the pre-railway era. South of Bridge 14 the canal widens at the site of Hatherlow Wharf where there were limekilns, whilst, on the towpath side, stood a hat works, one of many in the Stockport area. A pair of charmingly named aqueducts - Hatherlow and Burymewick respectively - carry the canal at rooftop level around the heavily suburbanised fringe of Romiley; steps leading down from the towpath to the streets they cross. The latter offers access to Chadkirk Chapel - see page 44. Oakwood Mill dates from the middle of the 19th century. Originally employed for cotton spinning, during the First World War it manufactured margarine. Nowadays, known as Romiley Board Mill, it produces cardboard products from recycled materials. *Continued overleaf:*

for details of facilities at Romiley and Marple turn to page 44

41

continued from page 41:

Hyde Bank Tunnel is just about wide enough for boats to squeeze past each other inside, though boaters of a nervous disposition may prefer to wait. The towpath, meanwhile, goes entertainingly over the top, itself bridged by the carriageway to Oakwood Hall. Ormerod Heyworth, builder of Oakwood Mill in the 1830s, erected this baronial pile on the opposite bank of the canal and lived there in considerable style. But like many an overblown property, time was unkind to it, and by the 1950s it had become a remand home for wayward girls. Having recently been extensively (not to say expensively) rebuilt, it was on the market for little shy of £2 million at the time of our previous research trip.

And then, achingly beautiful, the canal glides through woodland buttressed above the Goyt between Hyde Bank Tunnel's eastern portal and Marple Aqueduct. The supports date from a major earth slip in 1833. Bridge 15 heralds a narrow, retaining-walled section which was originally another tunnel.

Referring obliquely and somewhat misleadingly to it in his classic Victorian travel book *Wild Wales*, George Borrow rather memorably termed Marple Aqueduct a 'stupendous erection'. Completed in 1800, it bears the canal a hundred feet above a steeply wooded ravine carved by the impatient River Goyt*. Alongside, in juxtapositions reminiscent of Chirk on the Llangollen Canal, and Slateford on Scotland's Union Canal, stands an even loftier railway viaduct. From the right hand side of a Manchester to Marple Rose Hill train, slowing for the tightly curved junction, a bird's eye view of the aqueduct can be gained. Alternatively, from the aqueduct's western end, a stepped path - on which you may have to come to terms with vertigo - leads precipitously down past recently felled trees to offer a worm's eye view of the structure, this time allowing appreciation of its unusual cylindrical hollowed sections, so built to lessen the weight of masonry resting on support piers. Boatmen, it's said, were in the habit of burying their horses at the foot of the aqueduct. They must have been quite heavy to carry down.

*Up until the end of the 19th century the river was known as the Mersey downstream of the Goyt's confluence with the Etherow opposite Brabyns Park.

42

Over two hundred years old, Marple Aqueduct remains a monument to its Alfreton-born engineer, Benjamin Outram, who died from a stroke just five years after its completion, at the ridiculously early age of forty-one. We forget what a toll their tireless work took on these pioneering engineers of the industrial revolution, though the Butterley (Engineering) Company became his legacy.

Enormously satisfying for the spectator and participant alike, the sixteen Marple Locks raise the canal by over two hundred feet. They'll take the latter a leisurely two and a half hours to negotiate, but they took ten years to build, largely because the canal company was experiencing financial problems. This was far from unusual, for huge capital outlay was involved in advance of any profits which could come from completion. Indeed, for a number of years goods were laboriously transhipped from boats at either end of the flight onto a temporary tramway.

Highlight of the locks is undoubtedly Oldknow's Warehouse above Lock 9, a stunningly handsome, three storey stone structure now housing a suite of enviable offices. Samuel Oldknow - one of the main promoter's of the Peak Forest Canal - had a mill at Mellor, and bales of finished cotton were brought here by road for storage before onward transit along the canal. Bridge 18 is known as Posset Bridge, apparently because Oldknow promised the navvies a posset of ale apiece if they finished the bridge on schedule. History records that they did! The bridge incorporates separate horse and foot tunnels and, on the offside, an arch indicates a former arm to the foot of a bank of limekilns. Another short arm left the canal at this point and led to Hollins Mill. Green's coal boats traded here as late as 1959, one of the last commercial traffics on the canal. Take a moment before continuing along the canal to see 'Lock 17'*, a memorial to local historian Gordon Mills. The top four chambers of the flight must be unique in their setting alongside the neat front gardens of a street of suburban villas, and you can pass the time of day with the 'postie' on his round as you progress through the chambers. A series of extended side ponds have been transformed into duck-busy backwaters.

*A few yards to the east along Strines Road.

Macclesfield Canal

Marple is the meeting place of the Peak Forest Canal and its junior - by some thirty years - the Macclesfield Canal. Surveyed by Thomas Telford and engineered by William Crosley, the latter was promoted to provide a more direct route between Manchester and The Potteries than previously offered by the Bridgewater and Trent & Mersey canals, though its completion in 1831 coincided with the onset of the Railway Age and its full potential was never realised. Indeed, along with the Ashton and Peak Forest canals the Macclesfield was absorbed into its APM group of canals by what became known as the Great Central Railway, which, at the railway grouping of 1923, went on to become a constituent of the LNER.

Transport politics apart, Marple is a superbly photogenic junction, abounding in great sweeps of stonework and cobbles, about which there used to be a well kept air of confidence in accord with its strategic position on the Cheshire Ring, though on the occasion of our most recent research trip one had to admit that elements of the canal looked decidedly scruffy. One of the most obvious attractions is the Macclesfield Company's perishable goods warehouse, a durable stone building with an arched loading bay adjoining a former stop lock and a functionally stylish roving bridge which enabled horses to haul boats through the junction without the need to unhitch the towing rope. British Waterways unveiled plans for a residential redevelopment of Marple Wharf before reinventing themselves as the Canal & River Trust, but the new housing has been slow to materialise: 'on account of the economy', as Springsteen once sang. Scruffiness notwithstanding, there is so much to see that it is one of those locations which attracts more than its fair share of gongoozlers; all of whom have been urged, one imagines, on no account to miss Marple.

Between bridges 2 and 3 the canal narrows at the site of a former swing-bridge, and adjacent to this narrows stands a dilapidated corrugated-iron boathouse which recalls early leisure use of the Macclesfield Canal, encouraged (unusually for the period) by the Great Central Railway.

Goyt Mill, by Bridge 3, is one of the most impressive you'll encounter on the Cheshire Ring. Built early in the 20th century - predominantly of red brick with cream lining - it spun cotton imported through Manchester Docks and brought here by canal. Spinning ceased in the Sixties as the British textile industry declined in the face of Far Eastern competition. But like many of these stately northern mills, the huge building's echoing galleries now house an exotic variety of small businesses. It was to Goyt Mill that the last commercial traffic on the Macclesfield Canal - coal from Stoke-on-Trent - survived until 1957.

Marple Junction

Boathouse in decay

Romiley

Map 17

Suburban settlement with bustling main street best reached from Bridge 14 or Hyde Bank Tunnel.

Eating & Drinking

DUKE OF YORK - Stockport Road. Tel: 0161 406 9988. Comfortably appointed inn dating from 1786. Wide choice of food (ex Mon) and real ales. Two minutes walk from Bridge 14. SK6 3AN

PLATFORM ONE - Stockport Road. Tel: 0161 406 8686. *Good Beer Guide* listed pub beside the railway station. Wide choice of mainly locally brewed real ale and upstairs restaurant. SK6 4BN

QURASHI - Stockport Road. Tel: 0161 406 9000. Indian restaurant and take-away. SK6 3AN

SPREAD EAGLE - Hatherlow. Tel: 0161 494 5723. Nice refurbished 18th century pub easily accessed from Hatherlow Aqueduct. Open from 11.45am daily; food served lunch and evening weekdays and from noon throughout at weekends. SK6 3DR

IL PEPE NERO - Stockport Road. Tel: 0161 494 9979. Italian restaurant easily reached from Br. 14. SK6 3AA

Shopping

Good little shopping centre. Sainsbury's Local (on Compstall Road to east of station) and Thorley's family butcher (on Stockport Road west of station). Branch of Aldi adjacent Bridge 14.

Things to Do

CHADKIRK CHAPEL - Vale Road. Tel: 0161 480 1460. Open Sat & Sun & Bank Hol Mon afternoons. Ancient chapel and walled garden. Access from the towpath at Burymewick Aqueduct. SK6 3LD

Connections

BUSES - services 383/4 provide circular connections with Marple and Stockport. Tel: 0871 200 2233.

TRAINS - Local Northern trains provide connections with Marple, Hyde etc Tel: 03457 484950.

Marple

Maps 17/17A

Subtract the canal, and its setting on the cusp of the moors, and Marple would amount to little more than an innocuous suburban annex of Stockport. That said, the little town's position on the 'Cheshire Ring' makes it a popular overnight mooring point and it has some of the handiest and best facilities on this side of the circuit. Christopher Isherwood, author of *Goodbye to Berlin* - from which the musical *Cabaret* was derived - was born here.

Eating & Drinking

RING O' BELLS - Church Lane. Tel: 0161 427 2300. Robinson's pub by Bridge 2 on the Macclesfield Canal. Lunch and dinner daily. Accommodation. SK6 7AY

MARPLE SPICE - Stockport Road. Tel: 0161 427 9166. Indian restaurant near Bridge 18 PF Canal. SK6 6BJ

NO.48 - Stockport Road. Tel: 0161 427 1200. Spanish tapas bar/restaurant open daily ex Mon. SK6 6AB

THE RAILWAY - Stockport Road. Tel: 0161 427 2146. *GBG* listed pub adjacent to Rose Hill railway station and thus handy for the Middlewood Way. SK6 6EN

Two good micropubs: *Samuel Oldknow* on Market Street and *Beer Traders* on Stockport Road.

Shopping

Handy shopping centre (including a large Asda supermarket) within easy reach of bridges 2 or 18. Nice bakery called Archers on Hollins Lane: oh those corned beef and vegetable pies! Littlewoods butchers are worth seeking out on Church Lane.

Things to Do

REGENT CINEMA - old-fashioned, one-screen cinema adjacent Bridge 18. Tel: 0161 427 5951. SK6 6BJ

Connections

BUSES - frequent services to/from Stockport which is an interesting town (with a Hat Museum) to visit in its own right. Tel: 0871 200 2233.

TRAINS - station just down hill from Bridge 17. Services to/from Manchester, Romiley, Guide Bridge, New Mills etc. Tel: 03457 484950. Rose Hill station, just over half a mile west of the town centre, is the terminus of a half-hourly service from Manchester, but it also marks the northern end of the Middlewood Way (converted from its abandoned trackbed after closure in 1970) and has thus potential for walkers.

TAXIS - VIP. Tel: 0161 427 0777.

Marple Bridge

Maps 17/17A

Embarrassingly, it has taken Pearsons thirty years to discover the thriving little community grouped about the bridge which carries the road to Glossop across the Goyt. A sobering lesson to complacent guide book compilers that there is invariably something interesting just off the edge of the map.

Eating & Drinking

THE MIDLAND - Babyn's Brow. Tel: 0161 427 2370. Large refurbished pub with balcony over the Goyt. Named after the railway which arrived in Marple in 1876. Open daily from noon. SK6 5DT

DUTSONS - Town Street. Tel: 0161 484 5380. French inspired delicatessen/cafe and wine shop. SK6 5AG

TOWN STREET FRYER - Town Street Tel: 0161 449 7290. Fish & chip shop open lunchtimes and evenings (until 7.30pm) daily ex Sun & Mon. SK6 5AA.

MAPLE TREE - Town Street. Tel: 0161 484 5575. Chinese restaurant. SK6 5DS *That's Maple not Marple!*

Shopping

Dutsons (see above) excepted, not much use for provisions - you'll have to go into Marple itself for those - but charming for clothes, antiques, and a very good (if compact!) secondhand bookshop called Talisman Books - Tel: 0161 449 9271.

Things to Do

BRABYNS PARK - woodland and riverside walks.

KNOWN in some elevated circles as the *Upper* Peak Forest Canal, the route southwards from Marple, to the twin termini of Bugsworth and Whaley Bridge, is one of the most exhilarating on the system. Lockless, and punctuated but occasionally by moveable bridges, it is the setting as opposed to the canal itself which lifts this section out of the commonplace, like an otherwise ordinary tune rendered memorable by sumptuous orchestration. A vigorous landscape of fells, wind-bent woods, lonely stone cottages, railway viaducts and colossal mills places this particularly beautiful canal in an austere, northern mould, and canal explorers, whatever their means of propulsion, are blessed with a sense of privilege to be viewing the world from such a sublime perspective.

Your departure from Marple Junction may be subject to some delay as there is much to see. The sturdy stone house opposite the roving bridge was once part of James Jinks' boatbuilding yard. A former drydock has been transformed into a sunken garden. It became disused in the Thirties, the story being that it leaked so badly that Mrs Jinks' cellar was flooded every time a boat entered or left the dock. Adjoining the old boatyard, below an arm now used for moorings, is a bank of lime kilns which once had about them the look of medieval ruins. It is thought that Samuel Oldknow - who had his finger in every money-spinning pie between Manchester and Stockport - had the kilns so built to romanticise the view from Mellor Lodge, his home across the valley.

continued overleaf:

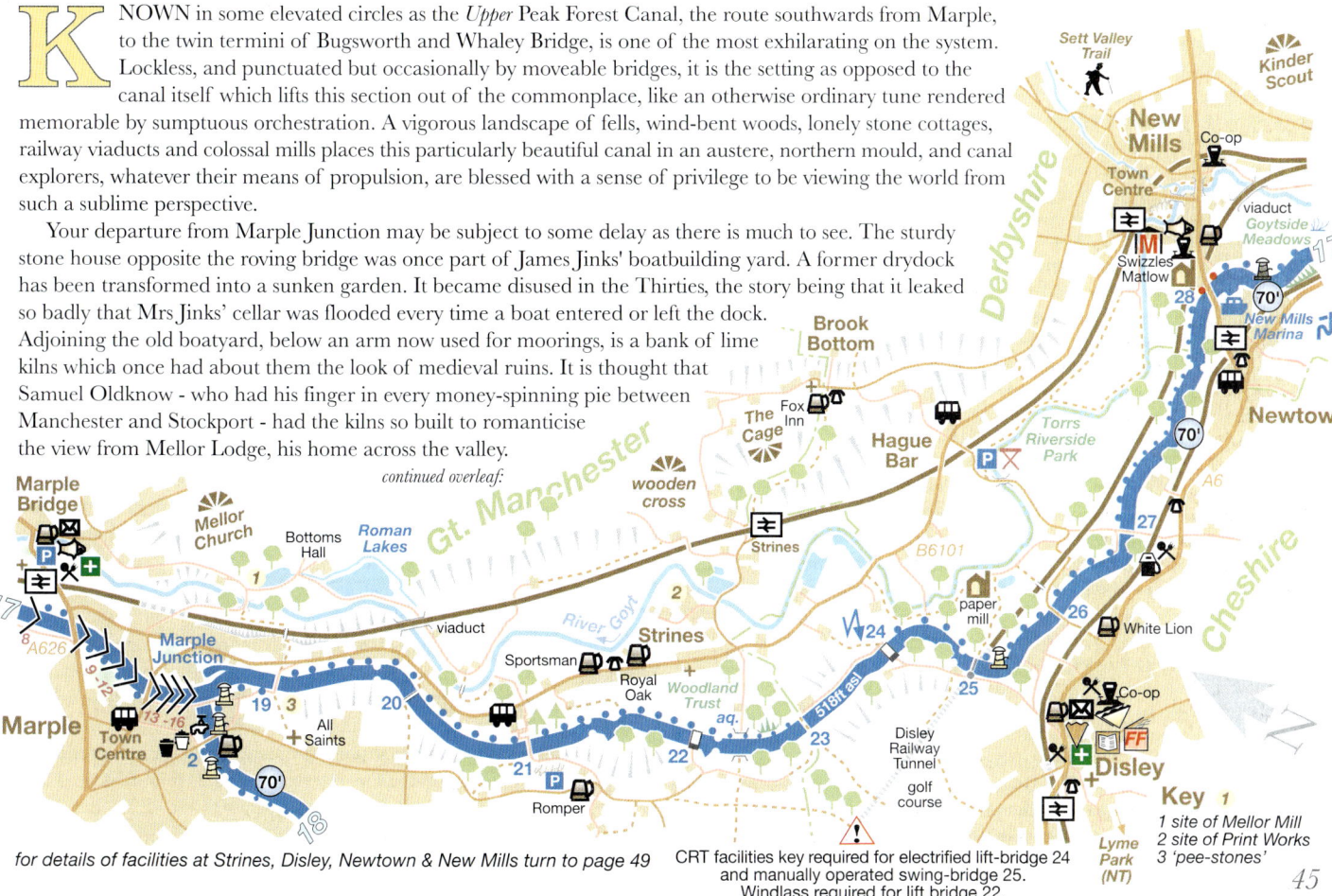

for details of facilities at Strines, Disley, Newtown & New Mills turn to page 49

CRT facilities key required for electrified lift-bridge 24 and manually operated swing-bridge 25. Windlass required for lift bridge 22.

Key
1 site of Mellor Mill
2 site of Print Works
3 'pee-stones'

continued from page 45:

Alongside Mellor Lodge stood Mellor Mill, a massive enterprise destroyed by fire in 1892. A group of industrial archaeologists are painstakingly restoring the wheel pit and associated water tunnels at the site of Mellor Mill which can easily (and entertainingly) be reached by footpath from Bridge 19.

The towpath changes sides at Bridge 19 which, unusually for the Peak Forest Canal, is constructed of brick. Oldknow's apprentices crossed the bridge on Sundays on their way from Bottoms Hall beside the river - where they were accommodated in conditions advanced for the day - to All Saints Church on the hillside to the west. On the way they'd make full use of two 'pee stones', thoughtfully provided - for what we'd now call a personal facility break - before the rigours of the service. If you walk up to All Saints now (encountering a capped-off mine shaft and those 'pee stones' en route) you'll see that only the tower of the earlier church (which Oldknow, with typical altruism had funded) remains, isolated in its extensive graveyard adjoining an imposing Gothic Revival replacement of 1880.

Keep your eyes peeled for a large wooden cross on the horizon to the east. It was erected in 1970 to commemorate a spot where John Wesley preached, and a service is held beside it every Easter. A landmark no longer visible was the lofty chimney of Strines Print Works, demolished in 2007 and replaced now by housing. The printing was on textiles rather than paper. The works had a wharf by Strines Aqueduct between bridges 22 and 23 for the unloading of raw materials and the loading of finished goods before the railway was built through the Goyt Valley. Incidentally, Strines station - which astonishingly, given its remote location, survived Beeching - has been suggested as the inspiration for the station in Edith Nesbit's *The Railway Children*. Certainly she came to stay with her step-sister at Mellor next door to a house known as 'Three Chimneys', intriguingly there is a scene in the book (though not the film) where a canal child throws stones at the railway children, whilst additionally Edward Ross, secretary of the Manchester, Sheffield & Lincolnshire Railway lived at Marple and may well have inspired the character of the 'old gentleman'.

Waterside Mill, visible as the canal curves round between bridges 24 and 25, can trace its history back over two centuries. Originally it belonged to Oldknow (who else!) but since the turn of the 19th century it has concentrated on the manufacture of paper, tissue being a speciality now. The Midland Railway's direct main line from Manchester to London emerges from Disley Tunnel (over two miles long) just below the canal. The portal is crowned by the MR's trademark wyvern and dated 1901. Bridges 21-23 & 26 sport good old fashioned British Waterways 'blue & yellow' number plaques, a nice custom to have revived.

New Mills Newtown is the largest settlement encountered by the Upper Peak Forest Canal, indeed, in most respects its development was brought about by the canal's construction. Here, in premises adapted from the old Brunswick cotton mill, is Swizzles-Matlow's confectionery works, purveyor of sherberty smells and origin of all the sticky substances you invariably discover abandoned in your children's pockets - even when they're in their twenties. There are two more mills and a foundry overlooking the canal from various angles which, between them, continue to evoke a very 'northern industrial' atmosphere. One of the mills is home now to Trafalgar Marine Services, fender makers and providers of rope. On the offside, New Mills Marina has been revitalised since this edition was originally published. It has become part of the Pridewater Estates group who also have marinas on the Grand Union Canal near Leighton Buzzard, the Coventry Canal near Lichfield, and the Birmingham & Fazeley Canal at Fazeley. On site holiday lets are also available here at New Mills in two of the boatyard's ancillary buildings.

Irritatingly, CRT give more prominence to long term moorings at Newtown than to the visiting variety. Surely more encouragement should be given for excursions to the fascinating little town of New Mills. But then they don't make money from visitor moorings, not directly anyway. As the canal twists around the hillside to reveal superb views of Kinder Scout - at 2,088 ft, the highest point in Derbyshire - there is access to Goytside Meadows, a local nature reserve whose trio of unimproved fields retain their delightfully euphonious nomenclature: Nice Eyes, Little Eyes, and Higher Flowery Croft.

17B PEAK FOREST CANAL Whaley Bridge & Bugsworth 2mls/0lks/1hr

THERE is going to be no escape from these glacially moulded hills and the canal knows it, but seems determined to enjoy itself nonetheless. Similarly obliged is the explorer, on foot, bicycled or afloat, for it would be churlish to be otherwise. Within the increasing confines of its valley, it is as if the Goyt has thrown a party and been overwhelmed by the number of attendees: the canal, two railways and the A6 trunk road, well on its way north from Luton to Carlisle. The railways, Midland to the east and London & North Western to the west, were absorbed at the 1923 grouping into the LMS, whereas the Peak Forest Canal, railway owned since as early as 1846, became part of the LNER. Co-ordinated transport is an elusive concept where the British psyche is concerned.

Furness Vale Marina creates regular tidal surges of pleasure craft. The long row of terraced cottages opposite would have provided homes for workers at the neighbouring calico printing works now used as an industrial estate. Raised above the level of the canal, trains to Buxton rattle by at hourly intervals; the signal box beside the level-crossing retains its London Midland Region 1950s maroon coloured enamel nameplate. On the opposite side of the valley the former Midland line hosts long trains of limestone quarried in the vicinity of Buxton, the modern equivalent of the narrowboats which once carried the same commodity so successfully on the canal.

Is this canal more conducive to happiness than any in the country? Pearson's, notorious for shifting (or perhaps more kindly, cyclic) allegiances, could certainly present an affirmative, if fleeting case. Bridge 34 is merely a footbridge now, but until relatively recently there was a lift-bridge here as well. Surplus to requirements, it has found a new lease of life on the Lichfield & Hatherton Canals. A pair of concrete road bridges introduces the junction of the original main line to Bugsworth Basin and the branch to Whaley Bridge; boat horses

continued overleaf:

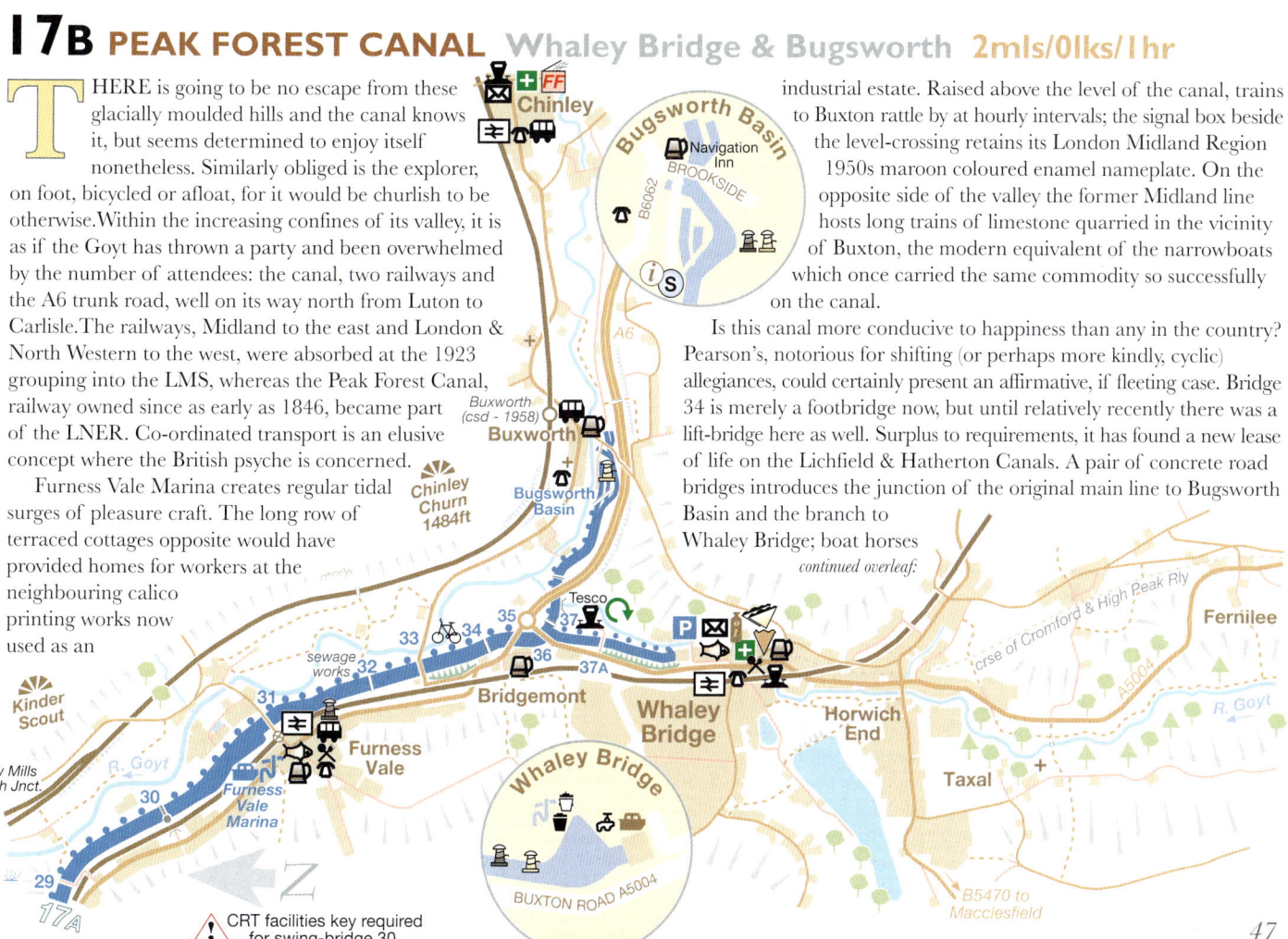

CRT facilities key required for swing-bridge 30.

47

continued from page 47:

gained access to the latter via a tunnel, and it is recorded that well broken-in animals would nonchalantly find their own way through the tunnel while their owners rope-hauled the boat across the junction by way of the footbridge.

An aqueduct carries the main line over the Goyt then it passes beneath the A6 before curving round to the reconstituted splendours of Bugsworth Basin. The Inland Waterways Protection Society may sound like some sort of boaters' mafia, but what they really are is a small, dedicated group of enthusiasts who have painstakingly restored Bugsworth Basins to something approaching the glory of their commercial heyday. The tireless efforts of the IWPS came to fruition with reopening of the basins in 1999. Unfortunately this proved a false dawn, and leakage brought about an abrupt closure of the basins to boating traffic, a hiatus lasting half a dozen years. Happily, however, the complex reopened for the second time at Easter, 2005. The event was marked by the symbolic departure of a horse-drawn narrowboat loaded with 16 tons of crushed limestone bound for Guide Bridge.

Bugsworth today enjoys Ancient Monument status and what the IWPS has achieved amounts to the successful excavation of an archaeological site of considerable importance. The complex comprises a series of transhipment arms and basins which radiate from a gauging stop overlooked by a wharfinger's house and canal office. A six mile tramway descended from the limestone quarries of Dove Holes. Loaded wagons ran down to Bugsworth by gravity, whilst empty ones were horse-drawn back. These trains were known as 'gangs', and often totalled twenty wagons at a time under the control - if that's not too precise a term - of a brakeman and his 'nipper', or youthful assistant; both of whom somewhat perilously rode the leading wagon. It comes as no surprise to learn that derailments were not exactly an unknown phenomenon on the rudimentary L shaped track. At Chapel-en-le-Frith there was an inclined plane which operated on the principle that empty wagons were hauled upwards by the weight of loaded ones travelling downhill. In the heyday of the basins perhaps twenty narrowboats a day would leave Bugsworth laden with limestone for Lancashire's burgeoning industries. The tramway ceased operating in the 1920s, though one of its wagons has found its way into the National Railway Museum at York. It would be difficult to exaggerate the reward to be gained from visiting Bugsworth, especially by boat in which you would have every opportunity to soak up the basin's latent atmosphere as dusk falls. The IWPS man an information centre alongside the entrance to the basins, and their patent enthusiasm for the site is contagious. Oh, and by the way, the natives took exception to the vulgarity of 'Bugsworth', considering Buxworth more polite.

Meanwhile, back on the 'branch' to Whaley Bridge, yet more private moorings signal the approach to this compact Derbyshire town. Shaded by woodlands, the canal runs parallel to the old, now by-passed A6 before terminating in a small, triangular basin dominated by a sizeable, stone-built, transhipment shed which formerly boasted three stories. An arm enters an archway in the centre of this handsome structure whilst, in the past, railway tracks were accommodated on either side to facilitate loading and unloading in sheltered conditions. Whereas the link at Bugsworth was by way of a fairly primitive tramway, the railway connection at Whaley Bridge, dating from 1831, was of a more sophisticated design. The Cromford & High Peak climbed right across the Peak District to link with the Cromford Canal, 33 miles to the south. Tantalisingly, the route was originally surveyed for the construction of a canal, but the Railway Age caught up with these heroic, not to say romantic proposals; more's the pity, when you contemplate what could have been one of Britain's most beautiful inland waterways.

For your part, you can stretch your legs along the course of the railway's steep incline which runs to the rear of the town's main street, and contemplate the profit and loss of historical chance. Canal termini are apt to have this effect on you, as if somehow mirroring life's own wasted opportunities. And where do we go from here, you ask? By train to Buxton perhaps - a scenic ride into the hills encountering the evocatively-named stations at Chapel-en-le-Frith and Dove Holes - or simply wend your way back along the gorgeous Peak Forest Canal, but not *too* quickly.

Disley
Map 17A

Amiable Cheshire village, even if it's hard to cross the A6 without leaving a limb behind. Nice walk to reach it along the lane from Bridge 25. Shadowing the Macclesfield Canal, the Gritstone Trail sets off for Kidsgrove from the railway station forecourt.

Eating & Drinking

DANDY COCK INN - Market Street. Tel: 01663 763712. Cosy little Robinson's pub. SK12 2AA
WHITE LION - Buxton Road. Tel: 01663 762800. *Good Beer Guide* listed pub on the A6 easily reached from Bridge 26. Welcoming, well-appointed and good for both real ale and food. SK12 2HA
SASSO - Market Street. Tel: 01663 765400. Italian restaurant. SK12 2AA
SAFFRON - Market Street. Tel: 01663 766016. Indian restaurant. SK12 2AA
RAM'S HEAD - Buxton Road West. Tel: 01663 767909. Large smartly refurbished former coaching house open from 11am; food from noon. SK12 2AE
Disley additionally boasts Chinese and pizza take-aways.

Shopping

Facilities include a Co-op (open 7am-10pm - 8pm on Suns), butcher, baker, pharmacy, post office, newsagent, and a tiny RBS bank.

Things to Do

LYME PARK - Tel: 01663 762023. Imposingly lavish National Trust property set in splendid grounds. If on foot, eschew the A6 and follow the lane from Disley station; though beware, it's a three mile hike from the canal. SK12 2NR

Connections

TRAINS - hourly Northern services to/from Stockport, Manchester and Whaley Bridge. Tel: 03457 484950.

Strines
Map 17A

Oddly oath-like community on the valley floor with a 'tin tabernacle' for a church. Good base camp for climbs up to Brook Bottom from which there are panoramic views over to Lyme Park's 'Cage'.

Eating & Drinking

SPORTSMAN - Strines Road. Tel: 0161 427 2888. *Good Beer Guide* listed pub on B6101 best accessed via footpath between bridges 21 and 22. SK6 7GE
FOX INN - Brook Bottom. Tel: 0161 427 1634. Cosy whitewashed Robinsons pub up in the hills. SK22 3AY

Connections

TRAINS - approx bi-hourly Northern services to/from New Mills and Manchester via Marple and Romiley. Tel: 03457 484950.

Newtown
Map 17A

A 19th century textile-based expansion of the North Derbyshire town of New Mills. Handy shops, pubs and a good fish & chip cafe. Trains as Disley.

New Mills
Map 17A

New Mills itself is a more than worthwhile ten minutes away on foot. Here, a delightful heritage centre (open daily ex Mon. Tel: 01663 746904 - SK22 3BN) interprets local history and there's a dramatic trail to be followed (over weirs, under viaducts and through a chasm) along the rivers Goyt and Sett. Plenty of shops (including a nice little secondhand bookshop tucked up High Street), banks and eating/drinking establishments. Trains to/from Manchester or via the picturesque Edale Valley to Sheffield.

Furness Vale
Map 17B

Useful pitstop with two pubs, a Chinese restaurant and fish & chips. Trains as Disley.

Whaley Bridge
Map 17B

A charming little Derbyshire town sheltering under the moors of Axe Edge. Make the canal terminus a convenient excuse to explore the surrounding countryside. Go up and see the reservoirs in the Goyt Valley, flooded after the First World War.

Eating & Drinking

BAILEYS - Market Street. Tel: 01663 734612. Cosy English restaurant. SK23 7AA
CASA DI PIZZA - Market Street. Tel: 01663 734333. Snug Italian. SK23 7AA
SHEPHERDS ARMS - Old Road. Tel: 01663 732384. Whitewashed Marston's pub just a few hundred yards south of the basin close to the course of the former C&HP Railway. *GBG* listed, stone-flagged floor and unspoilt within. SK23 7HR
Try also the Goyt Inn, slightly closer to the canal basin. Ditto fish & chips from The Fryery - Tel: 01633 732902.

Shopping

Plenty of small characterful shops where they are eager to pass the time of day with visiting canallers. Large 24 hour Tesco supermarket situated close to the junction with the Bugsworth Arm. Bike Factory for beleaguered towpath cyclists - Tel: 01663 735020.

Connections

TRAINS - Northern Trains hourly service to/from Buxton (a worth-while destination for an excursion ashore) Stockport and Manchester with stops at Furness Vale, Newtown and Disley. Tel: 03457 484950.

Bugsworth
Map 17B

NAVIGATION INN - adjacent Bugsworth Basin. Tel: 01663 732072. Cosy atmosphere and canalia. Food and accommodation. SK23 7NE Bus service 190 runs to/from Whaley Bridge and Chinley Mon-Sat.

Strines Aqueduct

Swinging Disley

Delivery at Bridgemont

New Mills

Hyde Bank West

Peak Forest Portfolio

Upper Peak Forest near Strines

Macclesfield Moments

Danes Moss

Macclesfield

Congleton

Bosley Locks

Bridge 76

Hall Green

51

QUICKLY establishing its character, the Macclesfield Canal gathers momentum, proceeding generally on a north-south axis, employing cuttings, embankments and aqueducts to avoid the need for locks. There is grace in its overbridges - one might even go as far as saying sagacity - and eloquence in its mileposts, counting the miles from Marple down to Hall Green (on the outskirts of Kidsgrove) and back. For the most part, the canal is rural, though High Lane, an elongated community strung out along the A6, introduces elements of suburbia. Mainly, though, one's gaze is directed eastwards where moorland rears up in handsome waves. The tower visible in the middle distance is known as "The Cage" and is thought to have been erected in Elizabethan times as a vantage point from which to watch the hunting of deer.

In its brief commercial heyday, the Macc's heaviest traffic was in coal mined in the vicinity of Higher Poynton. Much activity centred on Mount Vernon Wharf (Bridge 15) where a fleet of coal carrying narrowboats was maintained. A network of tramways and inclines connected the canal with Anson, Nelson and Park collieries, and business was brisk, at least until the arrival of the railways. Amongst the carriers was the famous name of Pickfords who could trace their origins back to an 18th century waggoner from Poynton. Happily an echo of canal trade survives in the form of the narrowboat *Alton*, which regularly plies between Macclesfield and Whaley Bridge/Buxworth selling coal, gas and diesel.

Nowadays the extensive basins at Higher Poynton are busy with private pleasure craft, but leisure boating is nothing new to the Macclesfield Canal. Its beauty was already appreciated by the Second World War when the North Cheshire Cruising Club (whose journal was endearingly known as "The Ditchcrawler") was established on an old arm at High Lane, purchased from the LNER in 1943. A peculiarity, then as now, was the use of boathouses - the aquatic equivalent of a lock-up garage.

South of Bridge 15 the canal widens into a shallow pool, probably brought about by a burst. Presumably it was cheaper to pay out compensation to the farmer than go to the trouble of repairing the canal bank. The bridges in this former colliery area differ in design from the ellipsoidal stone arches previously mentioned. Instead they are flat decked and designed to be easily raised in the event of subsidence

Greater Manchester

Cheshire

'The Cage'

17 5 mp 6 7 8 9 10 11 12 High Lane 70' aq. 13 Middlecale Farm Barlow House Farm Braidbar Boats mp 14 15 70' mp 16 aq. 17 18 19

Disley Railway Tunnel

Middlewood

55 Middlewood Way

Norbury Hollow

Higher Poynton

Boar's Head

Anson Engine Museum

Lyme View Marina

mp

Miners Arms

Wood Lanes

1 High Lane - closed 1970
2 Higher Poynton - closed 1970

High Lane Map 18

Strung out along the A6, with its interminable traffic, this suburban satellite of outer Stockport is of little attraction in its own right (Pevsner tersely considered the church 'poor') but canallers might find its facilities of use: pubs, fish & chips, fast food outlets, a Chinese restaurant, a well-stocked Spar with cash machine, post office/pharmacy, and a Conservative Club offering Northern Soul nights. Buses to Stockport.

Higher Poynton Map 18

The elaborate, orange-brick Boar's Head, and one or two other Victorian properties (39 Shrigley Road housed the former station-master) hint at Higher Poynton's industrial pedigree, and the station platforms remain intact, otherwise all is innocuously residential now; though the village does run to a football team in the Stockport District Sunday League.

Eating & Drinking

BOAR'S HEAD - Shrigley Road. Tel: 01625 876676. Homely pub popular with boaters and walkers. Famous for home made pies! SK12 1TE

COFFEE TAVERN - cosy cafe dating back as far as 1876 when it was opened by Lord Vernon, owner of the Poynton area coal mines, as a public reading room.
BAILEY'S TRADING POST - Lyme Road (adj. Br. 15) canalside dealer in boating accessories (fenders, gas, oil etc) plus canalia, refreshments and day boat hire. Tel: 01625 872277 - SK12 1TH

Things to Do

NELSON PIT VISITOR CENTRE - two minutes walk from Bridge 15. Interesting displays and interpretive material relating to Higher Poynton's coal mining past and the railway origins of the Middlewood Way. Leaflets (concerning everywhere in Cheshire *apart* from Higher Poynton if our experience is anything to go by) and toilet facilities. SK12 1TH
ANSON ENGINE MUSEUM - Anson Road (five minutes walk from Bridge 15). Tel: 01625 874426. Delightful museum celebrating the internal combustion engine, located on the site of Anson Pit, closed in 1926. Xanadu of fabled names - Gardner, Ruston, Hornsby, Tangye, National, Mirrlees etc etc. Open Fri & Sun,

Easter to October, 10am-4pm. SK12 1TD
Connections
BUSES - service P1 operates hourly, Mon-Sat to Hazel Grove via Poynton. Te: 0871 200 2233.

Wood Lanes Map 18

Once the location of Adlington Colliery, though you wouldn't know it now. A footpath leads idyllically eastwards up onto the slopes of Park Moor.

Eating & Drinking

LYME BREEZE - adjacent Bridge 18. Tel: 01625 871120. Little restaurant open for lunch Wed & Sun; lunch and dinner Thur, Fri & Sat. SK10 4PH
MINERS ARMS - Wood Lane North (near Bridge 18). Tel: 01625 872731. Food lunchtimes & evenings Mon-Thur and from noon throughout Fri-Sun (last orders 5pm on Sun). Bed & Breakfast. SK10 4PF
Connections
BUSES - service 392 operates Mon-Sat to Macclesfield. Tel: 0871 200 2233.

Whiteley Green

Slippery Steps

Reflections of Bridge 16

53

19 MACCLESFIELD CANAL Bollington 5mls/0lks/2hrs

PRIDE came before profit margins when the Swindells (unfortunate name) came to erect their two massive cotton mills at Bollington - Clarence and Adelphi - in the middle of the 19th century. With their flamboyant architectural embellishments they have the air of baronial country seats rather than places where hundreds toiled to earn a meagre living. In retirement, however, they lend Bollington a plausibility it might otherwise lack and act as stately guardians to this rarely less than exciting length of canal. Soaring above this atmospheric former textile town on a high, stone-laid embankment pierced by a lofty aqueduct - which acts like a sort of 19th century portcullis to Bollington - the canal traveller is treated to a series of eyecatching views, though it would be rewarding to have the vegetation cleared to reveal the embankment in its original rawness. Nature doesn't always know best.

North of Bollington the canal - closely accompanied by the trackbed of the Macclesfield, Bollington & Marple Railway, closed in 1970 but reinvented as the 'Middlewood Way' in 1985 - ducks in and out of woodland and alternates between cuttings and embankments. Bridges appear at frequent intervals and appear to 'staple' the canal together. Eastwards the Pennine escarpment keeps inspirational company with

the canal. Westwards, Alderley Edge ends abruptly on the Cheshire Plain. On a clear day Stockport and Manchester, are well defined on the horizon, whilst nearer at hand is Woodford Aerodrome. Two famous aircraft designs made their inaugural flights from here: the Lancaster bomber in 1941 and the Shackleton reconnaissance plane eight years later.

The parapet of Bridge 25 bears the inscription 'Lovers Leap June 1894' which apparently relates to the sobering story of a married man who'd turned out his wife and three children in preference for another woman. Public opinion was so against the erring individual that his home was set alight and a 'safe house' had to be found for the 'other woman'. Finally, feelings were running so high against the lovers that they made a suicide pact and drowned themselves in the canal. Obviously it was deeper in those days!

South of Bollington, Kerridge Drydock, formerly known as Endon Wharf, was a busy site where locally quarried stone was loaded onto boats. In the small hours of 29th February 1912 a large breach occurred in the canal opposite the dock, draining the

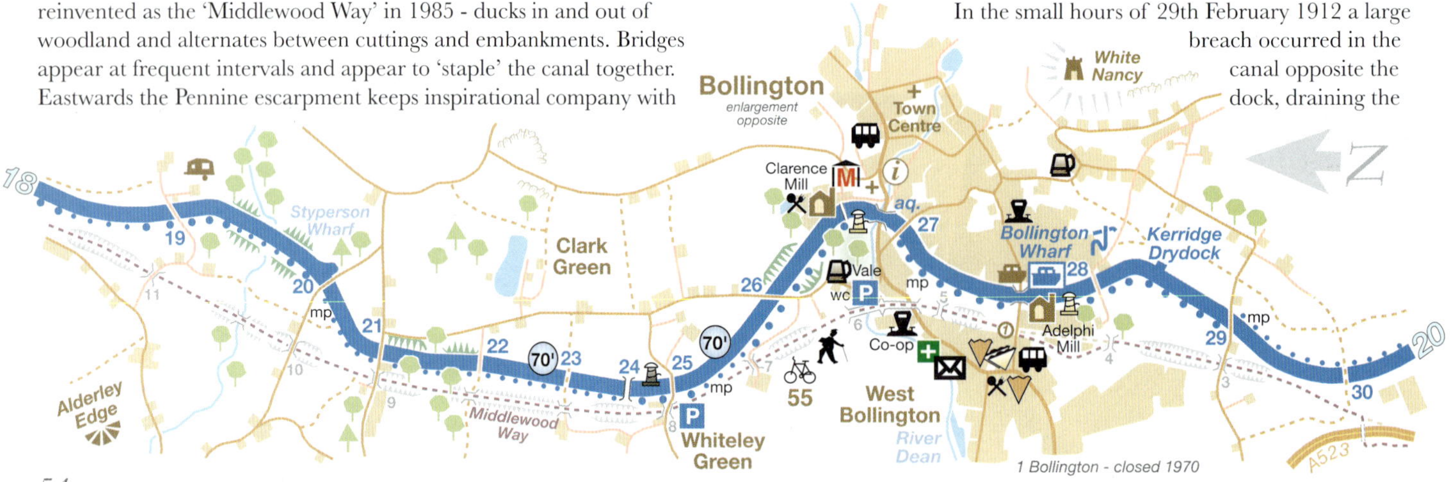

1 Bollington - closed 1970

54

whole twenty-mile pound between Bugsworth and Bosley Locks. Considerable damage was done to Bollington, and the town's gas works was rendered inoperable. 160 boat-loads of clay puddle were brought in to mend the breach. In stark contrast to the Dutton breach (Map 7) of a century later, it took three weeks as opposed to nine months to repair.

The towpath changes sides at Bridge 29, a typically attractive Macclesfield roving bridge - reminiscent of a neatly folded cardigan, wouldn't you agree? To the north-east the wooded slopes of Kerridge Hill dominate the skyline, topped by a whitewashed, sugarloaf-shaped stone monument. Built as a summerhouse to commemorate the Battle of Waterloo, it is known to all and sundry as 'White Nancy'. In 2012 'Nancy' was adorned with regal insignia to commemorate Queen Elizabeth II's Diamond Jubilee.

Bollington   Map 19

Nicknamed 'the Happy Valley' and boasting its own community radio station 'Canalside 102.8fm', the former cotton-spinning community of Bollington is as spick and span and as pretty as any hill town in Umbria - if you are going to 'drop anchor' anywhere on 'The Macc' this is the place to do so!

Eating & Drinking

BAY LEAF LOUNGE - Wellington Road. Tel: 01625 576465. Indian restaurant and take-away down from Adelphi Mill. SK10 5HT

BRISCOLA - Palmerston Street. Stylish and deservedly popular Italian restaurant. (Open Tue-Sat) Tel: 01625 573898. SK10 5PW

CAFE WATERSIDE - Clarence Mill. Tel: 01625 575563. Nicely appointed cafe for coffees, lunches and afternoon tea within Clarence Mill. SK10 5JZ

THE GREEN - High Street. Tel: 01625 576691. Stylish little cafe serving 'rustic food'. SK10 5PH.

LIME TREE - High Street. Tel: 01625 578182. Restaurant/wine bar serving 'meat from the family's farm'. Open 11am Tue-Sat, 10am Sun. SK10 5PH

TAPA - High Street. Tel: 01625 575058. 'Small plates & wine bar' open 5.30pm (4.30pm Sun). SK10 5PH

VALE INN - Adlington Road. Tel: 01625 575147. *Good Beer Guide* recommended pub owned by neighbouring Bollington Brewing Co. SK10 5JT

No.74 DELICATESSEN - Palmerston Street. Tel: 01625 573648. Charming deli/cafe. SK10 5PW

Clarence Mill

Shopping

Lots of nice individual shops whose owners seem genuinely interested in visitors 'off the cut'. In Bollington itself (amongst other useful outlets) you'll come upon Belfield's bakery, a deli, convenience store, newsagent and launderette, plus a branch of the alliteratively initialled J. J. Heathcote, the butchers up at the top end of Palmerston Street. Interiors outlet in Clarence Mill. Don't ignore West Bollington's facilities which include a Co-op convenience store with cash machine, Brassington's Bakery, *two* butchers (Barrow's and Heathcote's), and a pharmacy. The post office is on Wellington Road in West Bollington.

Things to Do

DISCOVERY CENTRE - Clarence Mill. Tel: 01625 572985. Open Wednesdays 1.30pm to 4pm; Saturdays & Sundays 11am to 4pm. Admission free. Splendid heritage centre operated by Bollington Civic Society. SK10 5JZ

BRIDGEND CENTRE - Palmerston Street. Tel: 01625 576311. Local information, internet access, and charity shop. Starting point for self-guided walks of the area. SK10 5PW.

Connections

BUSES - Arriva service 10 runs half-hourly (hourly Suns as 10A) to/from Macclesfield. High Peak service 392 connects hourly (not Suns) with Stockport. Tel: 0871 200 2233.

TAXIS - Bollington Black Cabs. Tel: 0797 145 1424.

20 MACCLESFIELD CANAL Macclesfield 5mls/0lks/2hrs

THE Macclesfield Canal looks down over the slate rooftops and terracotta chimney pots of the town which gave it its name. Here was the company's headquarters and their Brook Street Wharf, overlooked by a handsome mill which once belonged to Hovis the breadmakers. That the canal played an important part in transport to and from the mill can be seen from the arched loading bay at water level. Hovis transferred their milling activities to Trafford Park, Manchester, where ships could deliver imported grain and wheat direct to their door, but they still used this mill as a print works for their packaging and publicity material.

A cutting of high-sided stone retaining walls frames the canal's southern exit from Macclesfield. Perhaps Crosley, the canal's engineer, was already aware of construction techniques on the Liverpool & Manchester Railway, for there is a strong 'railway' character to this cutting. At Gurnett another cutting of this kind lies just north of an aqueduct which carries the canal above the road to Sutton Lane Ends

and the nascent River Bollin. East of here Tegg's Nose dominates the view, a former gritstone quarry which has become a country park in its retirement. Dry-stone walls criss-cross little fields climbing bravely up the hillsides, with here and there a knotty tentacle of stone terrace housing. This is Tunnicliffe Country. The famous wildlife artist was born at nearby Langley in 1901. Though best known as an illustrator of other author's works - Henry Williamson's *Tarka the Otter* for example - Charles Tunnicliffe also wrote and illustrated his own nature diaries, a number of which feature local scenes such as the canal reservoir at Bosley.

By swing-bridge No.47 (which boaters should ensure is returned to the closed position after use), a gnarled wood marks the perimeter of Danes Moss, a peat bog reminiscent of Whixall Moss on the Llangollen Canal. A footpath leads over the railway from Bridge 47 onto the moss, part of which is managed by the

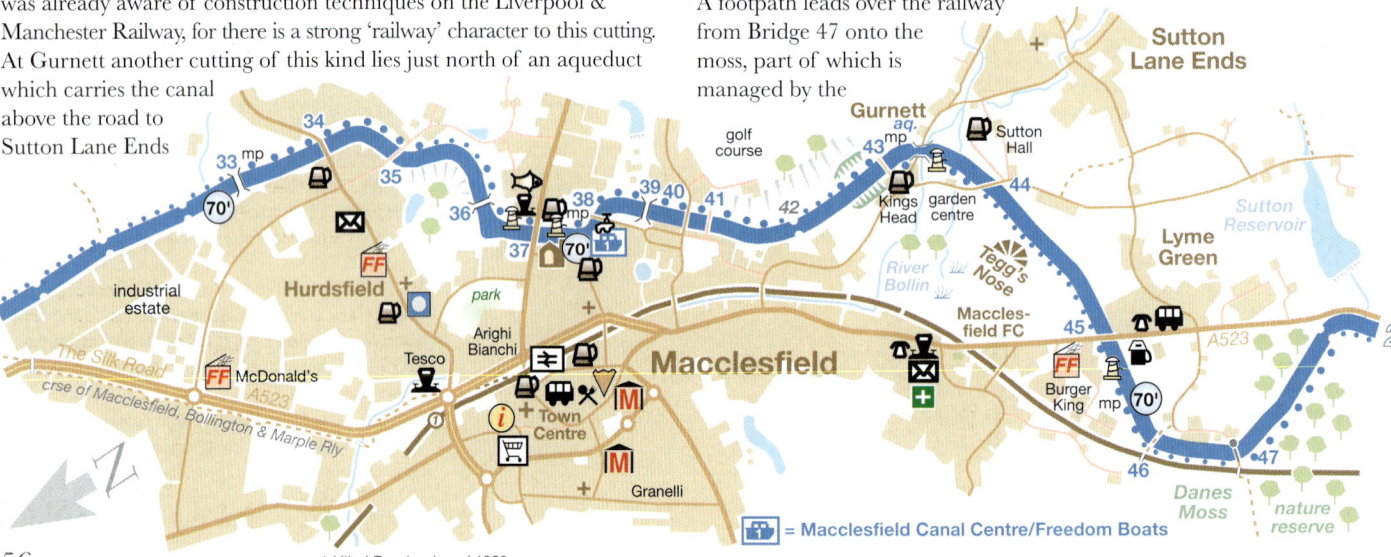

= Macclesfield Canal Centre/Freedom Boats

1 Hibel Road - closed 1960

Cheshire Wildlife Trust as a nature reserve, an agreeable wilderness of woodpeckers and willow warblers, dragonflies and damselflies, moths, butterflies and lizards. Visiting arrangements are informal but you are advised to remain on the path for your own safety!

Macclesfield
Map 20

The former silk weaving town of Macclesfield has some charming nooks and crannies, and some interesting museums which make it worth at least a semi-colon in the punctuation of any canal itinerary. Follow Buxton Road down from Bridge 37, past Arighi Bianchi's flamboyant furniture store, pass under the railway and you'll find yourself in Waters Green, where Wesley preached and which Tunnicliffe painted. Cobbled Church Street leads steeply up to the town centre and the fine, Greek Revival style Town Hall, adjoining which, the Tourist Information Office makes an excellent point of departure for further exploration of this old silk making town. Culturally, Macclesfield has become inextricably linked with the late 1970s post-punk band Joy Division, largely on account of its associations with Ian Curtis, the band's charismatic but ill-fated leader. Similarly, film buffs might be interested to learn that Macclesfield was used for the film version of James Hilton's *So Well Remembered* in 1947, starring John Mills.

Eating & Drinking
ABRUZZO - Mill Street. Tel: 01625 502937. Well-appointed Italian restaurant. SK11 6NR
CAFE BAR ARIGHI - The Silk Road. Tel: 01625 613333. Cafe/restaurant in stylish Italian furniture shop. Open Mon-Sat 9.30am-5pm, Sun noon-4pm. SK10 1LH
CHESHIRE GAP - 87 Mill Street. Vibrant deli/bakery/cafe. Tel: 01625 425806. SK11 6NN
FIVE CLOUDS - Market Place. Tel: 01625 429214. Microbrewery & bar, daily ex Sun. SK10 1EX
PUSS IN BOOTS - Buxton Road (canalside Bridge 37). Tel: 01625 263378. SK10 1NF

SALT BAR - Church Street. Tel: 01625 432221. Scandinavian cuisine. SK11 6LB
STAR LOUNGE - Mill Street. Tel: 01625 410600. Popular town centre Indian rest & t/a. SK11 6NR
TREACLE TAP - Sunderland Street. Tel: 01625 615938. *Good Beer Guide* listed continental style bar. SK11 6JL
WATERS GREEN TAVERN - Waters Green. Tel: 01625 422653. *GBG* listed town pub offering home-cooked lunches Mon-Sat. SK11 6LH
WHARF - Brook Street (to rear of Hovis Mill, access via Bridge 37 and Union Road). Tel: 01625 261879. *GBG* listed local to rear of marina. SK11 7AW

Shopping
Full facilities in the town centre 10 minutes walk from the canal. Cheshire Gap on Mill Street is an excellent delicatessen, whilst nearby a fishmongers called Cheshire Fish stands appropriately enough on Roe Street. Two providers of local delicacies are Marshall Spearing's (pies and sausages) on Park Green and Granelli's (ice cream) on Newton Street. Indoor market daily (ex Sun) outdoor markets on Tue, Fri & Sat. The Treacle Market (last Sunday of each month) is a venue for local food, antiques, books and artwork.

Things to Do
TOURIST INFORMATION - Town Hall, Market Street Tel: 01625 504114. SK10 1DX
HERITAGE CENTRE - Roe Street. Tel: 01625 613210. The story of silk and its development within Macclesfield housed in a former Sunday School. Plus local history. Shop and cafe. SK11 6UT
WEST PARK MUSEUM - Prestbury Road. Tel: 01625 613210. Fine and decorative arts, including the work of C. F. Tunnicliffe. SK10 3BJ

SILK MUSEUM & PARADISE MILL - Park Lane. Tel: 01625 612045. The Silk Museum is housed in Macclesfield's Victorian art school where Charles Tunnicliffe once studied. Paradise Mill housed a silk handloom weaving business. SK11 6TJ

Connections
TRAINS - 5 minutes downhill from Bridge 37. Virgin and Cross Country provide fast and frequent connections with Manchester and Stoke. Through trains to/from London. Northern stopping services run to/from Congleton. Tel: 03457 484950.
BUSES - bus station on Sunderland Street adjacent to rail station. Useful Arriva links with Lyme Green (service 14, hourly Mon-Sat) and Bollington (service 10 half-hourly Mon-Sat, hourly Sun). Tel: 0871 200 2233.
TAXIS - Silvertown. Tel: 01625 423333.

Gurnett/Sutton
Map 20

KING'S HEAD - Gurnett Aqueduct. Tel: 01625 423890. Good moorings make this a popular stop with boaters. Food and accommodation in quaint pub dating from late 17th Century. SK11 0HD
SUTTON HALL - Bullocks Lane (access via Bridge 44). Tel: 01625 253211. Impressive Brunning & Price conversion of a former manor house. Seven dining areas, terraces and gardens. Substantial menu and up to five real ales. *Good Beer Guide* listed. SK11 0HE

Oakgrove
Map 21

GAWSWORTH HALL - two miles west of Bridge 49. Tel: 01260 223456. Half-timbered house of considerable charm open to the public at Easter and from May to September. Concerts and special events. SK11 9RN

BOSLEY LOCKS is one of the most superbly engineered and magnificently located flights in the country; notable, for a narrow canal, in that both sets of gates to each stone chamber are mitred pairs. Once each chamber had a side pond, a water saving device which acted as a mini reservoir. When the lock emptied, half its contents would run into the side pond to be retained for half filling the chamber when it was next used. Theoretically this system halved the amount of water used by the flight. No one seems to recall exactly when the side ponds went out of use, but the Macclesfield Canal Society have hopes of resurrecting the side pond at Lock 4 for demonstration purposes.

The North Staffordshire Railway's Churnet Valley line (familiar to travellers on the Caldon Canal) crossed the canal between locks 11 and 12, the girder bridge which carried it still being in place, though barbed wire deters access. It wasn't the only railway in the vicinity. A 2ft 6ins gauge line ran from Thompstone's flour and corn mills in the neighbouring village of Bosley to a wharf alongside the canal below the bottom lock. The millers, apparently, would despatch their products by whichever means was cheaper at any given time, the Macclesfield Canal,

by that time being in the hands of the rival Great Central Railway. The canal crosses the River Dane on an imposing stone aqueduct, best (though probably illegally) appreciated from the path which descends to the valley floor from the north end of the neighbouring spill-weir. Blissfully peaceful and remote moorings are to be had below Lock 12.

North of Bosley the canal, at its summit level of 518ft, traverses the foothills of the Peak District. Between bridges 49 and 50 it occupies a shelf above a steeply sided valley reminiscent of a West Country combe. The woods are filled with chattering jays, whilst from spring through summer the air is heavy with the cloying scent of wild garlic.

Bridge 49 is an electrically operated swing bridge for which boaters will need a CRT facilities key

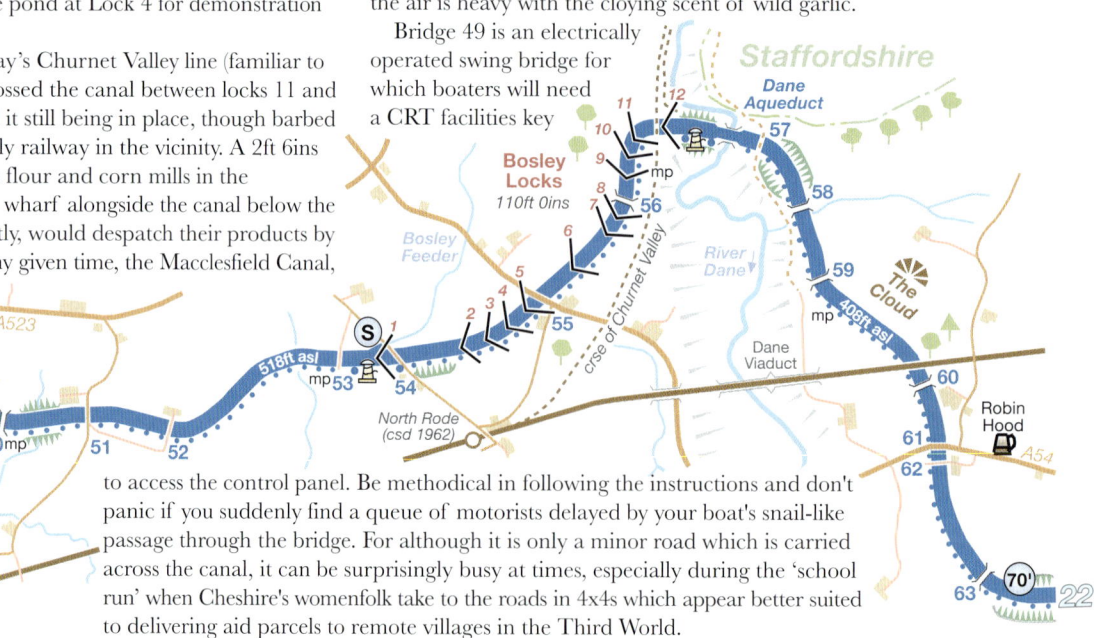

to access the control panel. Be methodical in following the instructions and don't panic if you suddenly find a queue of motorists delayed by your boat's snail-like passage through the bridge. For although it is only a minor road which is carried across the canal, it can be surprisingly busy at times, especially during the 'school run' when Cheshire's womenfolk take to the roads in 4x4s which appear better suited to delivering aid parcels to remote villages in the Third World.

22 MACCLESFIELD CANAL Congleton 5mls/0lks/2hrs

SOMETHING of a sea-change in the nature of the landscape occurs to the canal as it sashays past Congleton. Southwards the countryside grows softer and exfoliated as you head for the midlands. Northwards, rough-shaven hills with a Pennine hardness about them begin to close intimidatingly in on you.

Congleton's canalside is semi-detached and suburban, the old wharf having been assimilated almost seamlessly within a housing development. H. J. Lea Oakes organic animal feedstuffs mill overlooks an unusual gathering of transport modes by Bridge 75. A high embankment pierced by small aqueducts carries the canal over the steep-sided valley of Shaw Brook. A branch of the North Staffordshire Railway once threaded the valley and its trackbed has become a popular public footpath. A rail/canal transhipment dock known as Vaudrey's Wharf survives in water at Bridge 72. In the not too distant past we manoeuvred into its confines, imagining that we had a cargo to swap with the ghost trains. But word is now that it leaks and that there is no money available to have it relined, so it may well have to be filled in, which would be rather sad. Another memory relating to this length of the Macclesfield Canal concerns the gathering in for milking of a herd of cows by a cart horse. There appeared to be no farmer present whatsoever. As we cruised by, the horse came nonchalantly through a gate at the top of the field which sloped down to the canal. Clip-clopping around the field's perimeter, he got all the cows moving towards the gate, saw the last one safely through, and disappeared with the herd over the horizon: though it *could* have been two men in a pantomime horse!

By-roads lead from bridges 79 and 80 down to the village of Astbury, just under a mile to the west. The A34 doesn't do the village any favours, but the parish church of St Mary is remarkable and more than compensates. Pevsner considered it 'thrilling' which is about as excited as he ever got. Blink and you'll miss the canal's quiet passage over an aqueduct spanning another country lane.

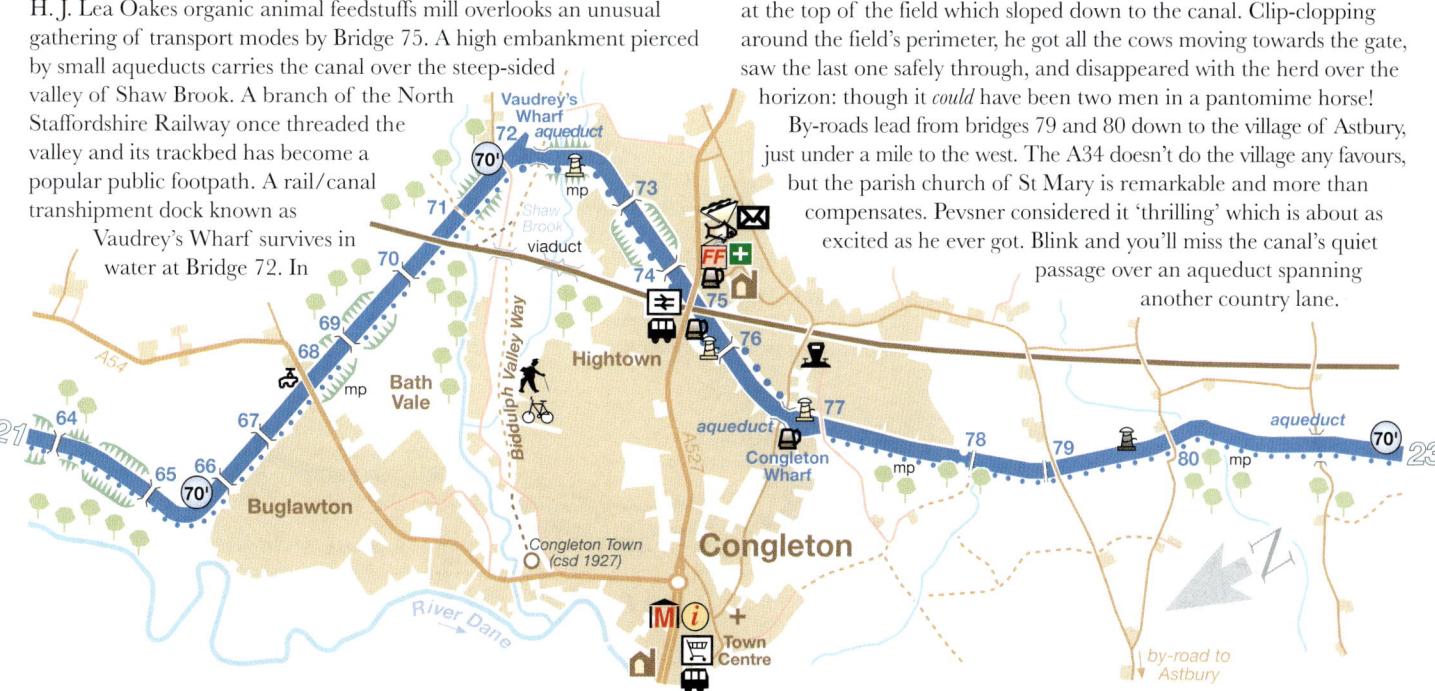

for details of facilities at Congleton turn to page 61

YOU don't want the Macclesfield to end. Like Rex Harrison - in *My Fair Lady* - you've grown accustomed to 'its' face. Those lucky perishers, smugly passing you on their way north, have all its beauty ahead of them. You've got, well, the Trent & Mersey; not an Audrey Hepburn of canals.

So make the most of it while you can. Play for time and go across the fields to the National Trust's incomparable Little Moreton Hall, or climb to the top of Mow (rhymes with cow) Cop. That castellated ruin you've watched growing closer for some time now, is Wilbraham's Folly. No, it's not some remnant of a medieval castle, it's an 18th century Angel of the North, a pure piece of whimsy plonked there for the entertainment of Squire Wilbraham of Rode Hall, a couple of miles to the west. In any case, the Cop's true claim to fame (other than that your author was once requested sternly to step down from it lest he endanger his life) is that the first open air camp meeting of the Primitive Methodist Revivalists was held on its heights one Sunday morning in May 1807.

As much as anything, you'll miss the Macc's milestones. So 'big ups' to the Macclesfield Canal Society who unearthed as many as they could (they were buried at the beginning of the Second World War to confound would-be invaders) or cut new ones from Kerridge gritstone to fill in the gaps.

Situated (until Beeching got his grubby hands on it) near Bridge 86, Mow Cop & Scholar Green railway station was immortalised (along with a number of other stations with canal connotations) by Flanders & Swann in their elegiac song *Slow Train*. The smooth lawns of Ramsdell Hall (c1760) sweep down to the canal bank. Untypically ornate cast iron railings (restored in 2008 to lovely effect) separate the towpath from a steep drop into the field below, so that the canal acts as a sort of ha-ha, or sunken wall, at the edge of the gardens. A mischievous image springs to mind of the gentry taking tea on the terrace and studiously ignoring the vulgar gaze of passing bargees - "Don't look, Daphne, don't look!"

The Macclesfield and Trent & Mersey canals are formally introduced at Hall Green as opposed to Hardings Wood Junction. The sequence of events which led to this anomaly is typical of the Trent & Mersey's paranoia in its dealings with potential

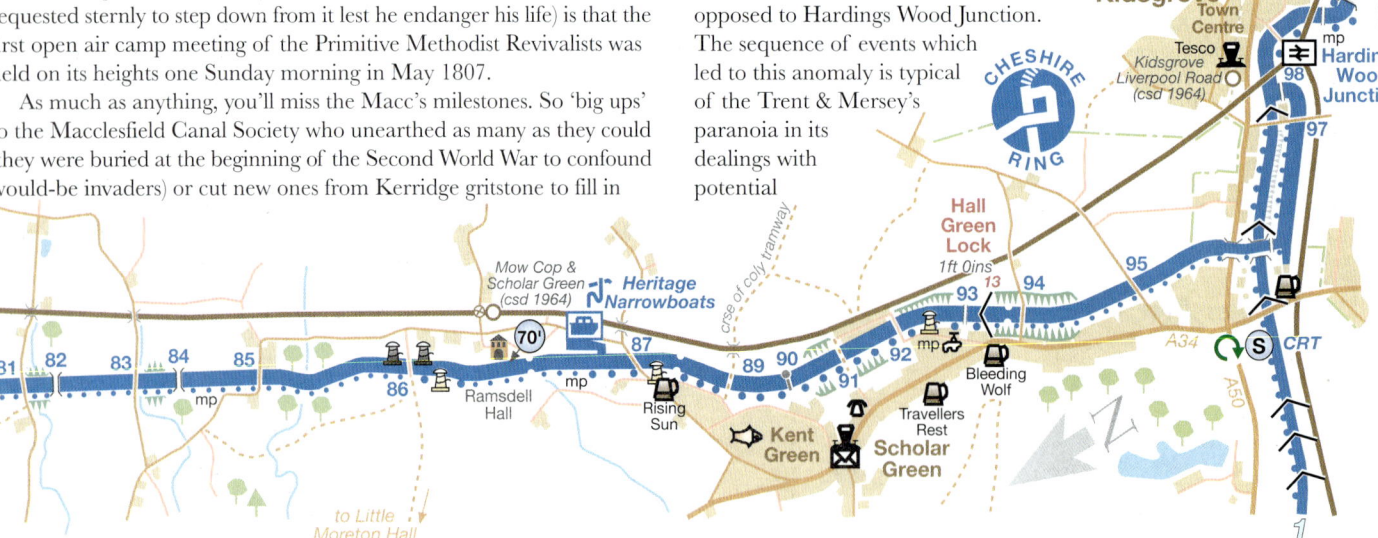

rivals. Far from welcoming trade which the new Macclesfield Canal might bring in its wake, the T&M saw only a threat to its established link between Manchester and The Potteries via Middlewich and Preston Brook. Feeling that they would be in a stronger position if the Macclesfield was not able to gain access to their main line, they built a connecting canal out to meet the newcomer at Hall Green. Here, both companies regarded each other in mutual mistrust across a 'Checkpoint Charlie' of paired stop locks, paired lock-keeper's cottages, and paired stable blocks. Neither in the railway age was their guard dropped, for the Macclesfield belonged to the Great Central, whereas the T&M was taken over by the North Staffordshire. Accidents of history to be embraced as you proceed across Pool Lock Aqueduct and back to Hardings Wood Junction, your hugely satisfying circumnavigation of Cheshire completed. Anyone for the South Pennine Ring? Bring it on!

Congleton   Map 22

Congleton has an unexpected zest about it, a level of activity which seems almost metropolitan to boaters, down from the cut with mud on their boots and a gauche unfamiliarity with road traffic. The best building in the town is the Town Hall, an imposing Flemish looking building which houses the excellent Tourist Information Centre. Lurking round the back is the town's museum which celebrates its bear-baiting past; the former mayor who ordered the execution of King Charles I; its former textile industries; cigar making; and its connections with Dutch soldiery during the Second World War.

Eating & Drinking
HIGHTOWN FISH & CHIPS - adjacent Bridge 75. Tel: 01260 276452. Open 11.30am-9pm daily ex Sun. CW12 3RJ
BEARTOWN TAP - Willow Street. Tel: 01260 270775. Welcoming local dispensing the ales of nearby Beartown Brewery. CW12 1RL
OATCAKES - Lawton Street. Tel: 01260 298040. Fresh filled oatcakes and pikelets, a rare Cheshire example of neighbouring North Staffs' staple diet. CW12 1RS
PURPLE PAKORA - Lawton Street. Tel: 01260 291110. Indian restaurant. CW12 1RP
QUEEN'S HEAD - adjacent Bridge 75, steps up from towpath. Tel: 01260 272546. *GBG* recommended. Food, beer and boules! Accommodation. CW12 3DE

YOUNG PRETENDER - Lawton Street. Tel: 01260 273277. Stylish bar related to the Treacle Tap in Macclesfield. Mon-Thur from 4pm; Fri-Sun from noon. Food served until 10pm. Great beer range. CW12 1RS

Shopping
Hightown offers a bakery, post office and pharmacy accessed via Bridge 75. The town centre is 10 to 15 minutes walk west of the canal, but there are bus connections - see below. Tuesdays and Saturdays are Market Days and there's a Morrisons supermarket by the bus station.

Things to Do
TOURIST INFORMATION CENTRE - Town Hall, High Street. Tel: 01260 271095. CW12 1BN
CONGLETON MUSUEM - Market Square. Tel: 01260 276360. Closed Mondays. Charming little repository of local history. Evidence that boating in Congleton goes back to 930AD. See how far you can get on the touch screen canal test. CW12 1ET

Connections
BUSES - half-hourly Mon-Sat to/from Town Centre from stop adjacent to Dog Lane Aqueduct. Ditto hourly from stop adjacent to railway station and Bridge 75. Tel: 0871 200 2233.
TRAINS - hourly Mon-Sat (limited Sun) Northern services to/from Stoke, Macclesfield and Manchester. Tel: 03457 484950.
TAXIS - Bear Town. Tel: 01260 291070.

Scholar Green   Map 23

Regrettably, the 'Mow Cop Killer Mile' - one of the most gruelling events on the running calendar - was cancelled in 2017 due to the 'spiralling costs of meeting health & safety regulations'. The summit is, however, well worth reaching at a more sedate pace. Views stretch from The Potteries to the Mersey and are as breathtaking as the climb.

Eating & Drinking
BLEEDING WOLF - access from Bridge 94. Tel: 01782 782272. Incongruous thatched thirties roadhouse on CAMRA's national inventory. Robinson's ales and good food. ST7 3BQ
RISING SUN - adjacent to Bridge 87. Tel: 01782 776235. Marston's pub within easy reach of the canal. Well thought of for its food. ST7 3JT
CHAN'S - Chinese-owned fish & chip shop accessible from Bridge 89. Tel: 01782 787083.

Shopping
Convenience store/post office counter on the A34.

Things to Do
LITTLE MORETON HALL - idyllic walk from Bridge 86. Tel: 01260 272018. Refreshments and NT shop. One of the most celebrated examples of a half-timbered house in England, Little Moreton Hall dates from the end of the 15th century. Top heavy, with leaning walls and sagging roofs, it seems like some vast, unstable doll's house. CW12 4SD

JUST when you begin to think that you must have been mad to leave the 'main line', and that nothing could be more tedious than the interminable and inscrutable industrial units of Trafford Park, the Bridgewater pulls a rabbit out of the hat. Some rabbit!

Brindley's original aqueduct over the Mersey & Irwell Navigation - deemed 'a canal in the air' in its day - was superseded in 1894 by an aqueduct which swung to facilitate the passage of sea-going vessels along the newly opened Ship Canal. Designed by Sir Edward Leader-Williams, and included in Robert Aickman's subjective 'Seven Wonders of the Waterways', it swings to this day, albeit much less frequently now. Canal cynosure it may be, but it is not the only notable edifice in the vicinity. Just uphill from the road bridge on the south bank of the Ship Canal stands Edward Welby Pugin's elaborate All Saints Church, paid for by the Trafford family, consecrated in 1868, and now fascinatingly occupied by the Conventual Franciscans.

The doyens of two transport modes briefly shake hands at Patricroft as James Brindley's Bridgewater Canal of 1761 passes beneath George Stephenson's recently electrified Liverpool & Manchester Railway of 1830.

Worsley is where it all began - the Canal Age that is. An era inspired by Francis Egerton, the 3rd Duke of Bridgewater's, serendipitous encounter - as part of his Grand Tour - with the Canal du Midi in 1753. Here a multi-level subterranean network of waterways was developed linking the canal with the coal faces of the Duke's mines. Boats known as 'starvationers', on account of the narrow, ribbed lines, were propelled from the coal face by artificial current. Surfacing at The Delph - through a now embowered portal - they were towed three abreast by pairs of horses to Manchester. Modern transportation had been set in train, and it is only mildly ironic that Worsley should be sliced in half by the M62 Transpennine motorway today. Westwards from Worsley, the canal threads its way between old mines and mosses to meet with the mighty Leeds & Liverpool at Leigh. The Royal Horticultural Society are restoring a 150 acre garden at Worsley New Hall.

Between Water's Meeting (Map 13) and Worsley, and Marsland Green (Map 25) and Leigh the Bridgewater Canal's towpath has been fully upgraded. Elsewhere, surfaced with colliery waste, it's prone to morass like muddiness after heavy rain.

Trafford Centre

Asda

Ashburton Road

All Saints

Barton upon Irwell

13

Barton Swing Aqueduct

Worsley CC

Patricroft

82

70'

Parrin Lane

Monton Green (csd - 1969)

MANCHESTER SHIP CANAL

M602

A57

12

1

B5211

M60 (M62)

13

Worsley

Worsley Dry Docks

The Delph

golf course

RHS Bridgewater Gardens (2019)

25

Boothshall

Bridgewater Marina

for details of facilities at Worsley turn to page 64

25 BRIDGEWATER CANAL Astley Green & Leigh 5mls/0lks/2hrs

ONE would dearly liked to have known this austere landscape when it still worked for a living; before, if you'll accept the simile, it acquired the melancholy sonority of a brass band. Old colliery basins abound, and it bears remembering that barges were continuing to carry coal from mines in the vicinity to Trafford Park power station as recently as 1972. Would that they still were; would that colliery engines were still puffing and panting across the flat fields with trains of protesting wagons on tortuous tracks to basin-side tipplers where their contents would be summarily upended and shot into the black holds of waiting 'Wiganers'; would that miners still trudged home along the towpath, too tired to bother avoiding the boat-horses' dung, too blackened by coal dust for their faces to be recognised in the South Lancashire dusk.

At Astley Green a skeletal colliery headstock recalls the past, whilst the community itself reassures you that civilisation does exist. Half an hour's stroll southwards along 'Rindle Road' will bring you to the wastes of Chat Moss where

Stephenson was forced to lay his railway tracks on floating bundles of faggots to prevent the new iron way from sinking into the bog.

Ceremoniously opened by George V, no less, in 1934, the East Lancs Road lays claim to be Britain's first inter-city highway; as historically significant, then, as the Bridgewater Canal or the Liverpool & Manchester Railway. Certainly it cuts an impressive swathe as it skews across the canal on a bridge redolent of its Art Deco period. Its next point of call to the west goes by the euphonious name of Lately Common.

At Bedford, a harbinger of Leigh, coal gave way to cotton. Vast textile mills - some, such as those occupied by Leigh Spinners, manufacturers of carpets and synthetic turf, still operative - rear up alongside the canal, their domed towers glinting like mosques in what passes for daylight hereabouts. Modern housing has replaced the weaving sheds which stood alongside Butt's Bridge, but interpretive boards, erected at frequent intervals as part of the Bridgewater Way initiative, and as yet unmarred by graffiti, provide insight into the history of the mills, not least the one on Mather Lane which could boast the first reinforced concrete floor in the world.

At Leigh Bridge the Bridgewater gives way imperceptibly to the Leeds & Liverpool. Imperceptibly in that there is no obvious junction, though through travellers do find themselves exchanging the capitalist waters of the Peel Group for the charitable waters of the Canal & River Trust, with all the ideological clashes such a step-change implies.

for details of facilities at Astley and Leigh turn to page 64

63

Worsley
Map 24

Half-timbered houses and broad swards of grass ensure that Worsley attracts land and water-based visitors alike. St Mark's Church chimes thirteen at 1pm! There are lots of pleasant strolls to be had beside the orange-tinged canal, whilst here and there interpretive boards outline the village's busy commercial past. Indeed, with a bit of spare fat on your Cheshire Ring schedule, Worsley makes a worthwhile detour up from Water's Meeting, perhaps for an overnight stay.

Eating & Drinking
BRIDGEWATER HOTEL - Barton Road (overlooking canal). Tel: 0161 794 6206. Large *Good Beer Guide* listed 'Fayre & Square' pub open at noon. M28 2PD
GEORGE'S - Barton Road. Tel: 0161 794 5444. Stylish restaurant overlooking canal and named after the architect Sir George Gilbert Scott. Brunch from 10am, food thereafter throughout from noon. M28 2PD
TUNG FONG - Worsley Road. Chinese restaurant. Opens 5.30pm. Tel: 0161 794 5331. M28 2NL
THE MOORINGS - Boothstown Marina. Tel: 0161 703 8895. Another 'Fayre & Square' outlet. M28 1YB

Shopping
Shopping facilities are somewhat surprisingly limited to a small newsagents. Presumably the natives all go to the ghastly Trafford Centre now.

Connections
BUSES - service 33 runs frequently to/from Wigan and Manchester. Tel: 0871 200 2233

Astley Green
Map 25

Idyllic post-industrial community overlooked by its colliery head gear and a Methodist chapel - services Sunday at 11am, 'all welcome'. A nice quiet place to moor between Manchester and Leigh.

Eating & Drinking
BAMBU - Higher Green Lane. Tel: 01942 876497. Chinese/fish & chips from 4.30pm ex Mon. M29 7JB
OLD BOATHOUSE - Higher Green Lane (canalside Astley Bridge). Tel: 01942 883300. Comfortably refurbished pub with waterside terrace, archive photographs of old Astley Green and a good choice of food. Lunch and dinner Mon-Thur (5pm), and throughout from noon Fri-Sun. M29 7JB

Things to Do
ASTLEY GREEN COLLIERY MUSEUM is open Sunday, Tuesday and Thursday afternoons throughout the year. The mine closed in 1970 but the massive, 100ft high headgear haunts the horizon still, the last of its iconic kind in Lancashire. Another survival is the Yates & Thom winding engine, all 3,300hp of it - just like a Deltic! Demonstration colliery railway and customised wagons on sale for railway modellers. M29 7JB

Connections
BUSES - services 551, 683/4 run to/from Leigh. Tel: 0871 200 2233.

Leigh
Map 25

It's hard for guide book compilers of a certain age not to hear Eddie Waring's elongated vowel version of 'Leigh' (as part of one of his trenchant Rugby League commentaries) when they think of this surprisingly substantial town, which has a longer pedigree than its prevailing image of a mill and mine community implies. The noble town hall dates from 1907 and is, according to the local publicity handout, "a dignified expression of civic prosperity and pride," and we wouldn't quibble. Elsewhere terracotta and Portland stone evince an Edwardian determination to compete with Wigan, Bolton and St Helens. James Hilton, author of *Goodbye Mr Chips*, *Lost Horizon* and *Random Harvest* - all novels turned into notable movies in the golden age of Hollywood - was born in Leigh. *So Well Remembered*, one of his less well known novels, was set in a mill town, but when they came to make the film they used Macclesfield for the outdoor scenes. Leigh Sports Village is an impressive development to the south of the canal, with a stadium capable of seating ten thousand spectators. On a more practical level, the town centre and all its facilities lies just north of Leigh Bridge, by which there are ample moorings.

Eating & Drinking
WATERSIDE INN - Twist Lane (canalside Leigh Bridge). Tel: 01942 605005. Purists may bristle that the Greene King brand should manifest itself so far from its traditional Suffolk homeland, but this pub in a converted canalside warehouse provides handy sustenance from 11am daily. WN7 4DB
THE THOMAS BURKE - Leigh Road (beyond Market Place). Tel: 01942 685640. Wetherspoon conversion of former Hippodrome Theatre named after 'the Lancashire Caruso' who habitually went through his singing exercises on the canal towpath. WN7 1QR
There are Frankie & Benny, Nandos and Chinese restaurants all canalside at the Loom retail park east of Leigh Bridge, but sadly no mooring facilities are provided.

Shopping
A massive Tesco 'Extra' supermarket stands canalside in the new Loom retail park east of Leigh Bridge, but, as observed above, no facility is offered for mooring, so you'll just have to trudge back to Leigh Bridge with your laden bags. Indoor market adjoining the Spinning Gate shopping centre. By Butt's Bridge, Bedford, Fishers family butcher does a nice line in pies and black pudding.

Connections
BUSES - X34 provides a robust link with Manchester. Tel: 0870 200 2233.
TAXIS - Z Carz. Tel: 01942 677107.

South Pennine Ring

Todmorden

REMARKABLY transformed from dereliction to navigable standards in 2002, the Rochdale Canal has not been taken to the bosom of the boating classes as avidly as an inland waterway miracle of such magnitude deserves. Perhaps this has something to do with the unnerving reality of eighty-three wide beam locks in thirty miles; though, as so often in life, popularity is not necessarily proportionately linked with value and worth, and one should not be deflected from so character-building an odyssey by middling hardships, bureaucratic or otherwise. Furthermore the Canal & River Trust no longer insist on the pre-booking of passages through locks 81-65, though volunteer lock keepers are happy to provide assistance where required, and their invariably cheerful and invaluable help can be pre-booked by telephoning: 0303 040 4040.

Engineered by Jessop - with contributions by Rennie and Crosley - and opened throughout in 1804, the Rochdale Canal departs Manchester's Piccadilly Basin and immediately lives up to its Alpine reputation by negotiating a couple of locks in quick succession. Refurbished textile mills provide piquant contrast with the futuristically brash upstarts of New Islington, as Ancoats is now politely known. All the mills have histories too

long for the telling here, but they repay close inspection, and to whet your appetite a plaque on the wall of Brownsfield Mill by Bridge 90 notes that it played a part in the nascent years of the aircraft industry. A carved inscription above the canalside entrance to Royal Mill informs us that the King and Queen 'popped-in' on 19th November 1942. Tell-tale signs in the offside brickwork show where subterranean arms once led into the mills. A new towpath bridge boomerangs across the canal with 'Cast No Shadow' amusingly reflected in the waters beneath it.

Part re-dug from old arms, part brand new, New Islington Marina is something that central Manchester has been crying out for ever since the Cheshire Ring was formed in 1974. Full boating facilities are accessible from the Rochdale Canal and secure visitor moorings are obtainable too. Incorporated into an urban park, you could almost imagine you were in some Spanish resort were it not for the neighbouring mills and the distantly brooding horizon of Saddleworth Moor.

So far so good! Then, abruptly, beyond Bridge 88 the wand of regeneration appears to lose its magic as smart apartments give way to the unimaginative inner city housing of the 1970s and a towpath liberally coated in Canada Goose droppings.

*Figures relate to Rochdale Canal east of Piccadilly Basin

= New Islington Marina

New Islington or old Ancoats?

Engels would be appalled to see how money - or more pertinently the lack of it - still divides the nation. At least the reinvigorated canal goes someway towards alleviating the gloom, manifesting our inland waterways' capacity for simultaneously regenerating both bricks and mortar and people's innate sense of well-being.

Soon the canal is encountering more locks, from which there is little respite until Failsworth is reached. Another fine example of mill architecture broods over Lock 79. Temples to Mammon they may have been, but, unlike succeeding places of work, they brooked no architectural argument. Locks 80-78 recall the long lost activity of coal mining, and a little bit further on the pound between locks 77 and 76 has been deepened to make adjustments for subsidence. Paradoxically, a little further on, in the vicinity of Newton Heath, are a number of shallow sections retained for the benefit of rare flora and fauna.

Barbed-wire topped brick walls mask wastegrounds which underline Manchester's industrial decline, a diminution further evidenced by the survival of just one workshop from the once vast Park Works which paralleled the canal between bridges 83A and 83. The engineering firm of Mather & Platt developed a 50 acre site from 1900 onwards, having purchased, at a literally knock-down price, the machinery hall of the

Victoria Mill

Royal Mill

Paris Exhibition of the same year, which was transported across the Channel, through the Irish Sea and up the Manchester Ship Canal for re-erection here.

More housing borders the canal as it makes its way through Newton Heath where the prefab canalside library garishly illustrates much of the neighbourhood's industrial past. Newton Heath was the original home of one of the world's most famous football clubs - Manchester United - who developed from a railway workers amateur team. Canalside street names such as Silk and Millwright echo a lost era of textile activity, though several of the mills live on. By Lock 67, Failsworth Home Guard resolutely continue to protect us against the possibility of invasion.

FAILSWORTH HOME GUARD CLUB

LOCK 65 at Failsworth marks the end (or, for westbound boaters, the beginning) of those gruelling locks. Travelling eastwards boaters will be grateful for a bit of a rest from the blessed things. Time to pay attention to the surrounding 'scenery' perhaps. The Rochdale Canal's commercial zenith was shortlived on account of the onset of the Railway Age, but at least its wide-beam barges provided a capacious logistical resource for a multitude of waterside mills and sundry other industries which sprang up along the route of the canal. Failsworth is particularly rich in mill architecture, to which the reborn canal lends fresh perspective. On the opening day in 1804 - when heavy frosts threatened to put a damper on celebrations by freezing the canal entirely - the band of the First Battalion of Manchester & Salford Independent Volunteers boarded one of the first vessels to pass along the canal and enlivened its passage by regaling bystanders with popular tunes of the day.

Between Failsworth and the M60 the canal passes under the new Metrolink line to Oldham and traverses a green corridor, an unlikely oasis much favoured by fishermen of all ages and genders. Wriggling through a cat's cradle of motorways and dual-carriageways, the canal enters an uninspiring zone of wastegrounds and modern warehouses, one bearing the good old engineering name of Widdop (though this one deals in giftware) and another, Voith. When the motorway was built, supporters for restoration of the Rochdale Canal managed to win a landmark battle in Parliament to have a culvert of navigable dimensions included in the scheme so as to safeguard the canal's future.

Oldham's skyline of chimneys, spires and high-rises rears up to the east but won't necessarily inspire a desire for closer acquaintance in many psyches. Middleton Junction no longer lives up to its railway origins, the branch lines to Oldham and Middleton having been closed in 1958 and 1964 respectively, but it remains the location of a signal box enchantingly entitled Vitriol Works. Incidentally, the Oldham branch included the fearsome Werneth Incline, at 1 in 27 by far the steepest passenger line in Britain.

The handsome Greengate Brewery of J. W. Lees & Co - family brewers in the vicinity since 1828 - lies just the other side of the railway. One imagines the beer in the fetchingly named Railway & Linnet couldn't be much fresher! Grimshaw Lane Lift Bridge (75A) is operated electrically and lifts, unusually, on a horizontal plane.

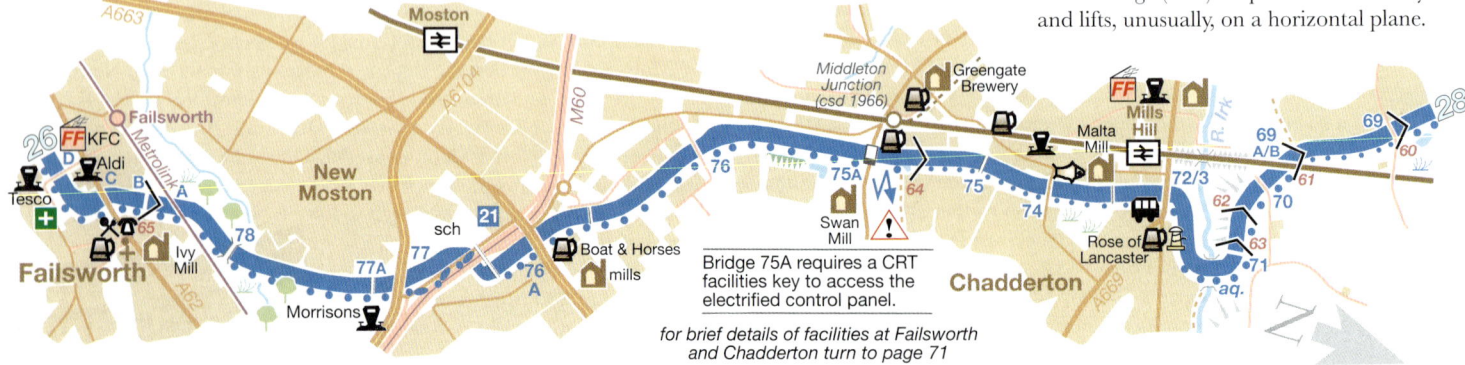

Bridge 75A requires a CRT facilities key to access the electrified control panel.

for brief details of facilities at Failsworth and Chadderton turn to page 71

It is overlooked by the massive Swan Mill, an outlier of many still extant in Oldham. It was erected for the Swan Cotton Spinning Company in 1875 and continued in production until 1959, notwithstanding a serious fire in 1922. These days it provides commodious warehousing, and, it goes without saying, the workforce no longer report for duty in clogs and shawls.

By Rochdale Canal standards, Lock 64 is oddly isolated. Eastbound it precedes another traditionally built up area of mills and housing on the road between Middleton and Oldham. Vinegary aromas issue from Sarsons' works to the rear of Malta Mill, used these days for document storage. Chadderton was associated for many years with aircraft production. Amongst other iconic designs constructed at the AVRO plant were Lancasters, Shackletons and Vulcans. Good moorings are available between the Rose of Lancaster pub and River Irk Aqueduct. These are especially significant when travelling westbound. Boaters need an earlyish start if they are to reach Manchester in one fell swoop. Mind you, the geese and donkeys in the field opposite the pub are often only too happy to oblige with a wake up call.

The railway (Manchester & Leeds, engineered by George Stephenson and opened in 1841, just 37 years after the canal)) thrusts its confident way across the valley on a hefty embankment. More demurely, the canal follows the contours before accepting that there's nothing for it but to sound the bugle and go over the top. Formerly four track, the railway has been reduced to two, suffering almost the same fate as the canal in terms of freight carried; nowadays - global warming notwithstanding - it is naturally the M62 which bears the brunt of the Trans Pennine logistics business. The last cargo to traverse the whole of the Rochdale is thought to have been the Calder Carrying Co.'s barge *Thomas*, consigned from Manchester Docks to Dewsbury with dried fruit for use as a dye in September 1937. Appropriate, perhaps, that the dying industry should provide the last breath of life on a dying canal.

Lees Brewery, Middleton Junction

Irk Aqueduct

Railway bridge near Chadderton

28 ROCHDALE CANAL Slattocks, Castleton & Rochdale 4mls/11lks/5hrs

SLATTOCKS is said to be a corruption of 'South Locks'; and this is easily believed, they still talk 'laike that oop 'ere'. Above Lock 54 are what one might term the first truly 'salubrious' visitor moorings since leaving Manchester, ten hours back. By gum you will feel you have earned a rest if you've made it this far.

One of the biggest challenges facing the engineers involved in re-opening the Rochdale Canal was to pass under the M62 motorway. Fortunately, a culvert for a farm track provided a possible answer. Thus a deviation from the canal's original course has been essayed at significantly less cost than once anticipated. At the same time, Lock 53 has been re-sited to the south of the motorway. A 'floating towpath' provides walkers and cyclists on 'Route 66' with a novel means of negotiating the restricted space available. Incidentally, the course of the Heywood Branch of the canal, lies beneath the motorway.

Eastbound, the landscape begins to drop hints of the glorious hill country to come. Moorland provides a theatrical backdrop to Rochdale, breezes ruffle high grasses in rough pasture grazed by ponies. On Tandle Hill radicals are said to have drilled prior to Peterloo. But just when you begin to think you have shaken off Manchester entirely,

the Rochdale Canal is reintroduced to industrialisation at Castleton where steam locomotives are occasionally glimpsed making their way to or from the preserved East Lancashire Railway based at Bury.

The canal restorationists were faced with a major road blockage on the southern edge of Rochdale. Resourceful engineering, and a determination previously unknown in the inland waterways' confrontation with road developments, won the day, though at the expense of the towpath user who is faced with a minor detour.

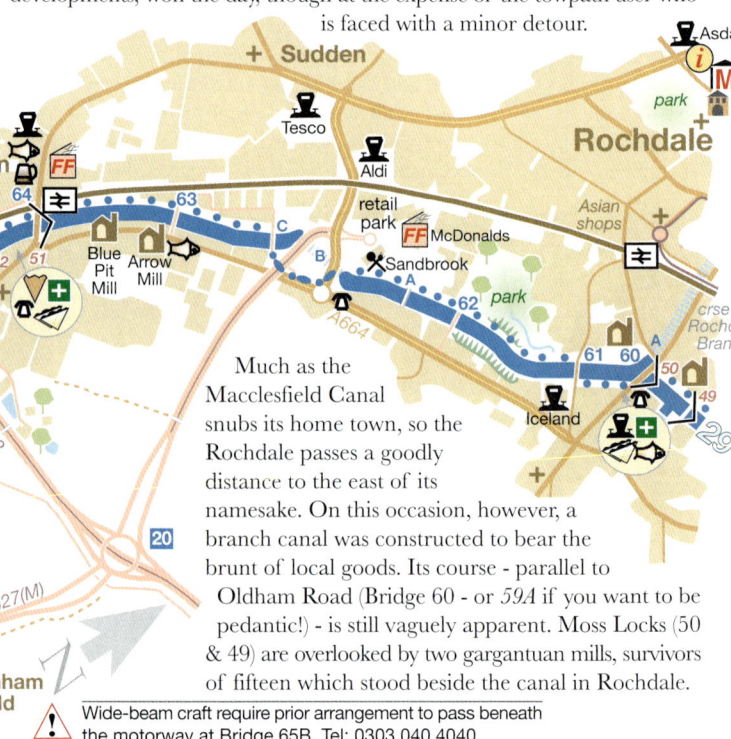

Much as the Macclesfield Canal snubs its home town, so the Rochdale passes a goodly distance to the east of its namesake. On this occasion, however, a branch canal was constructed to bear the brunt of local goods. Its course - parallel to Oldham Road (Bridge 60 - or *59A* if you want to be pedantic!) - is still vaguely apparent. Moss Locks (50 & 49) are overlooked by two gargantuan mills, survivors of fifteen which stood beside the canal in Rochdale.

⚠ Wide-beam craft require prior arrangement to pass beneath the motorway at Bridge 65B. Tel: 0303 040 4040.

Failsworth
Map 27

Boaters are not encouraged to tarry, but Failsworth is not without historical interest: viz the eponymous 'pole' and statue of local literary figure Ben Brierley. There's a large canalside Tesco, an Aldi, pharmacy, several pubs, and a KFC.

Chadderton
Map 27

Outlier of Oldham which could once claim fifty cotton mills. Birthplace of the particle physicist and television broadcaster Professor Brian Cox.

Eating & Drinking
ROSE OF LANCASTER - Haigh Lane. Tel: 0161 624 3031. A *Good Beer Guide* listed Lees 'all-day' pub (there's one on every corner hereabouts) open from 11.30 onwards (noon on Sundays). Canalside patio at street level. Water by arrangement. OL1 2TQ

Shopping
Small convenience store west of canal between bridges 74 and 75; Tesco Express to west of railway station accessed from Bridge 72/3.

Connections
TRAINS - local services to/from Manchester, Rochdale etc from Mills Hill station adjacent Bridge 72/3. Tel: 03457 484950.
BUSES - services 59/64 run to Manchester (via the amenable little town of Middleton) and Oldham via Chadderton town centre. Tel: 0871 200 2233.

Castleton
Map 28

A suburb of Rochdale known in times gone by for its blue clay. The lofty spire of St Martin's church provides a landmark alongside the canal. Consecrated in 1862, it suffered so much from dry rot that it had to be closed in 1991, but has since been re-opened with English Heritage funding. Handy railway station for towpath walkers; ditto fish & chip shop.

Slattocks
Map 28

On 21st May 1866 an Irish labourer was brutally murdered with a crowbar outside the Hopwood Arms. The perpetrator, eighteen years old James Burrows, was convicted and publically hung at Salford Gaol. Thirty thousand members of the public attended.

Eating & Drinking
Two canalside pubs vie for your custom (Ship Inn - Tel: 0161 643 5871, and the aforementioned Hopwood Arms - Tel: 01706 359807) but if you've still enough energy left we would recommend the charmingly isolated Tandle Hill Tavern (Tel: 01706 345297 - M24 2SD) just beyond the motorway bridge at the bottom of the map, a worthwhile and enjoyable twenty minutes walk from the canal.

Shopping
Londis convenience store at garage.

Connections
BUSES - service 17 operates at frequent intervals to/from Manchester and Rochdale. Tel: 0871 200 2233.

Rochdale
Maps 28/29

We bang on about boaters needing secure town moorings, and Rochdale's no worse than any other in this respect, but imagine the difference it would make to the community's coffers if you *could* moor snugly here overnight, or leave the boat securely for a visit to this never less than entertaining town. In the absence of such facilities (think 'boat parks' as opposed to *car* parks) we would advocate catching the train from Littleborough, because it would be a missed opportunity to be in this part of the world and not pay one's respects to the home town of Gracie Fields. The Town Hall is jaw-dropping in its municipal ambition and majesty and would do justice to many a grander town. The story goes that Adolf Hitler coveted it, and would have shipped it back to Berlin if he'd won!

Eating & Drinking
THE SANDBROOK - Sandbrook Way (adjacent Bridge 62B. Tel: 01706 719660. Stonehouse Pizza & Carvery open from 8am-11pm. OL11 1RY
BAUM - Toad Lane. Tel: 01706 352186. *Good Beer Guide* listed pub beside the Rochdale Pioneers Museum. Up to eight real ales on tap and stylish food. OL12 0NU
CLOCK TOWER - Town Hall. Tel: 01706 924772. Open 9am-4pm, Mon-Fri. Breakfasts, lunches and afternoon teas. A council run restaurant well worth patronising to surreptitiously assimilate the Town Hall's Gormenghastian interior. OL16 1AB

Shopping
The Exchange and Wheatsheaf shopping malls host all the predictable High Street names. More lively and authentic is the traditional covered market, open daily save for Sundays. Effervescent area of Asian shops on Milkstone Road south-west of railway station.

Things to Do
TOUCHSTONES - The Esplanade. Tel: 01706 924492. Art gallery, local studies centre and cafe. Open Tue-Sat 10am-5pm. OL16 1AQ
PIONEERS MUSEUM - Toad Lane. Tel: 01706 524920. Celebration of the Rochdale Pioneers who founded the Co-operative movement in 1844. Open Tue-Sat 10am-5pm. OL12 0NU
FIRE SERVICE MUSEUM - Maclure Road (adjacent railway station). Tel: 01706 341219. Open Fridays and first Sunday in the month. Housed in handsome fire station dated 1933. OL11 1DN

Connections
TRAINS - Northern Trains links along the Calder Valley and to/from Manchester. Tel: 03457 484950.
TRAMS - Metrolink service to/from Manchester via Oldham. Tel: 0161 205 2000.
TAXIS - Streamline. Tel: 01706 644104.

ROCHDALE'S burgeoning suburbs are beginning to spill across the canal, and the new Metrolink (whose nifty, Viennese, yellow and grey trams zip across the canal by swing-bridge 58A) will probably encourage even more development once the economy gets back on an even keel; if you'll excuse the half-hearted pun. And with the houses come supermarkets, fast food outlets, garden centres and all the other accumulated baggage of the 21st century. Take a deep breath, this will soon be behind you (always assuming you're eastbound). Indeed, once you've popped through the hole of Bridge 56 (coming face to face with a goat in the process) the scene is rural; well, quasi-rural. And then, running briefly upon an embankment parallel to the railway (note the L&YR boundary posts) the views widen out left and right and an elated sort of vindication is yours for all the effort put in since Piccadilly Basin. Firgrove Mill's horizontal single tandem condensing engine, surplus to requirements in 1970, found salvation in Manchester's Museum of Science & Industry.

Would that they took on obsolete guide-book compilers too.

The canal passes through a shallow shale cutting before reaching a handsome group of former weavers' cottages and a mill refurbished with apartments standing either side of an astonishing early 17th century mansion called Clegg Hall. It was completed circa 1618 by one Theophilus Asheton and during its long and not always illustrious career became a public house known variously as the Hare & Hounds or Black Sloven. Shotblasted clean, it is wuthering no longer, exuding rather the raw pink sheen of a baby's bottom. To the west the flailing arms of a group of wind turbines provide a landmark on Scout Moor. To the north lies Brown Wardle Hill. The imposing clock tower in the middle distance belongs to Birch Hill Hospital, erected as a workhouse in 1877, and a well known 'spike' in its day amongst 'gentlemen of the road'.

Sheep silhouetted on the summit of Cleggswood Hill usher in Littleborough, western terminus of the restored section of the Rochdale Canal from 1996 until the re-opening thence to Manchester six years later. The outskirts turn a somewhat blank face as the canal skirts its eastern periphery, but it's a welcoming little town where solace can

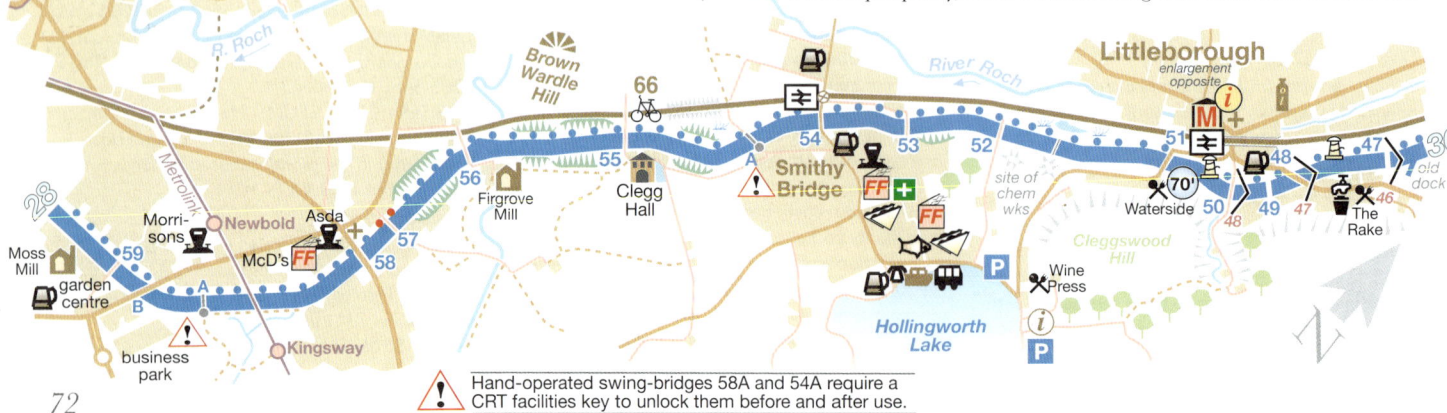

Hand-operated swing-bridges 58A and 54A require a CRT facilities key to unlock them before and after use.

be found along the shelves of Kelsall's excellent secondhand bookshop, or, if literature is not your bent, by simply watching your underwear go round and round in the launderette. If it's fun you want, make your way to Hollingworth Lake, a 118 acre canal reservoir apparently under the misguided impression that it's a seaside resort complete with amusement arcades, ice cream kiosks, fish & chip bars, and enthusiastic crowds of day-trippers with rolled-up trousers and tucked-in skirts. Earlier generations of Rochdale and Oldham textile workers knew it as 'the weavers' seaport' and walked up from the stations at Smithy Bridge and Littleborough in

their droves. Captain Webb is reputed to have used the lake in his training for the first cross Channel swim. Swimming, alas, is no longer encouraged, but water-sports are, and Hollingworth is also part of a Country Park.

East of Littleborough the canal begins to climb again, ascending to its short, but dramatic summit. By Benthouse Lock (46) an angled dock, complete with the stump of a former crane, was apparently employed in loading stone quarried on Blackstone Edge. The lock derives its name from a 17th century property which overlooks the canal as it broadens into a pool (Map 30).

Littleborough Map 29

A traveller in the days before the canal or railway came to Littleborough, found it "a very desirable retreat when it is found impossible to ascend the mountains, during the continuance of the howling storm." That tradition continues, for there is still an inclination to sit tight in this typically friendly little 'Lancashire' town, waiting for the skies to clear before tackling the summit. Even in more clement weather you'll find plenty of excuses for delaying departure.

Eating & Drinking

THE WATERSIDE INN - Inghams Lane (Bridge 51). Tel: 01706 376250. Delightfully atmospheric canalside bar and restaurant. OL15 0AY
THE RAKE - Blackstone Edge Old Road. Tel: 01706 379689. Tapas bar/restaurant with its own micro-brewery. OL15 0JX
WINE PRESS - Hollingworth Lake. Tel: 01706 378168. Bar/restaurant with lake views. OL15 0AZ
MR THOMAS'S - Hollingworth Lake. Tel: 01706 373731. Eat in or takeaway fish & chips overlooking the lakeside resort. OL15 0DQ
RED LION - Halifax Road. Tel: 01706 378195. *Good Beer Guide* listed pub between the railway and the canal. OL15 0HB

THE SUMMIT - cosy pub adjacent to western end of canal summit. Tel: 01706 379500. Thwaites beer. Food served 9am-8pm daily. OL15 9QX

Shopping

Surprisingly extensive facilities including: Co-op adjacent to railway station and Sainsbury's 'Local' on Hare Hill Road. Some good butchers, bakers and greengrocers: the dilemma of choosing between a pie from the Littleborough Deli or the Village Bakery next door is solved by purchasing one of each. Kelsall's antiquarian bookshop (Tel: 01706 370244)

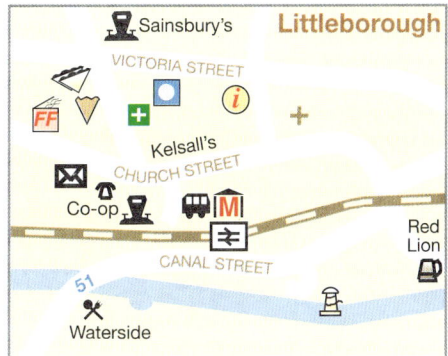

has an extensive collection of new and s/h books on the locality. Launderette on Victoria Street. Community Market second Sunday of the month.

Things to Do

COACH HOUSE HERITAGE CENTRE - Tel: 01706 378481. Closed Mondays. Open 11am-4pm daily (from 1pm Sun). Community hub with exhibitions of Littleborough's history and Tourist Information. OL15 9AE
HISTORY CENTRE - local history displays located on the eastbound platform of the railway station and open weekends and bank holidays. Tel: 01706 838385.
HOLLINGWORTH LAKE VISITOR CENTRE - Tel: 01706 373421. Information about the surrounding countryside. Cafe and shop. OL15 0AQ
HOLLINGWORTH LAKE ACTIVITY CENTRE - Hollingworth Lake. Tel: 01706 370499. Rowing boat hire and trip boat operation, sailing, windsurfing, canoeing etc. OL15 0DQ

Connections

TRAINS - half-hourly Northern link with Todmorden and Rochdale (easiest way of visiting for boaters); hourly with Walsden. Tel: 03457 484950.
BUSES - services 452/5 run up to Hollingworth Lake. 590 parallels the railway service over the summit into Yorkshire. Tel: 0871 200 2233.
TAXIS - Newline & Phoenix Tel: 01706 378000.

WITH the strippling River Roch* in tow, the Rochdale Canal gets quickly 'stuck in', as a series of locks carry it up past the old Rock Nook cotton mills. Fothergill's once manufactured khaki drill for the army. Now they produce high performance textiles for the aerospace and defence industries. The mill by Bridge 43 is occupied by a company dealing in theatrical scenery and prop hire. Did you notice how the Roch is taken over the railway by the western portal of Summit Tunnel in a sinuous iron trough? Summit was the eastern terminus of Rochdale Corporation's tramway network. In the 1920s heyday of urban tramways, it would have been feasible (though downright eccentric) to have journeyed in a sequence of electric tramcars all the way across Lancashire from Summit to Liverpool. Na that would 'ave bin summat!

A quiet sense of pride and achievement marks your attainment of the summit, an oxygen-challenged six hundred feet above sea level. You'd be forgiven for thinking that this is the highest canal pound in the land, but that honour goes to the South Pennine Ring's other Transpennine canal, the Huddersfield Narrow (see Map 38), which is forty-five feet higher; though the majority of that is spent in a tunnel, so perhaps it shouldn't count!

Right here, right now, however, your affinity lies with the Rochdale as its summit pound slithers through a writhing landscape of haphazardly scattered hillocks spliced by towering electricity pylons. For a ravishing interlude of tantalising brevity you find yourself hob-nobbing with the gods. At Bridge 41 you pass from Greater Manchester into Calderdale, aka West Yorkshire. Mind you the boundary has moved back and forth down the years. Pity the poor away side pitched against Todmorden Borough FC at their canalside Bellhome ground, playing not only the home team, but their craggy, fog-bedevilled environment to boot. Worth a two goal start by any reckoning!

What goes up, must come down. One feels like Moses returning from Mount Sinai. A delusion not discouraged by the lovely ladies behind the counter in Grandma Pollard's, who regard you as if you might be carrying news of supplementary commandments. 'Thy shall put copious salt on thy chips'. Locks with euphonious nomenclature like Winterbutlee (30) and Nip Square (29) lower the canal out of the hills into a land of mill chimneys and north-light weaving sheds. Trains hoot peremptorily as they enter Summit Tunnel. Short intervening pounds have a predilection for spilling over into reedy margins. By Bridge 35 St Peter's Victorian church is possessed of a thin black spire and a musical chime. Post WWII prefabs line the canal by Lock 26.

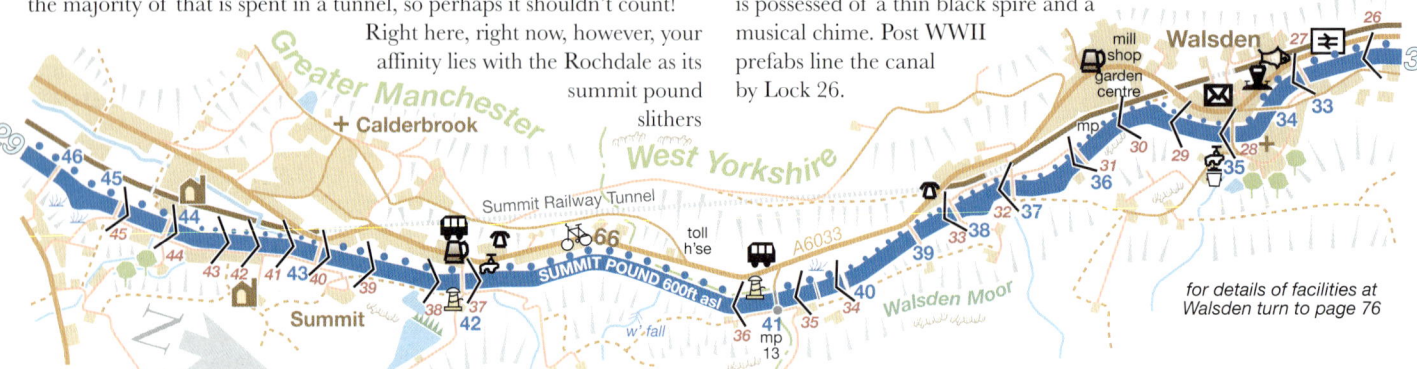

for details of facilities at Walsden turn to page 76

* pronounced 'Roach', as in the fish.

GAUXHOLME boasts one of the most visually satisfying canalscapes in the kingdom, its most potent image the canal's passage beneath the Gothically inspired, cast iron railway bridge with battlemented abutments - particularly photogenic from the neighbouring hillside - easily reached (assuming you've a Sherpa with you) from Bridge 31.

Past cobbled cul-de-sacs still strung with washing lines, you continue your descent. Such is the contiguity of the locks, that you move at a snail's pace - even by canalling standards - but it hardly matters, such is the drama of the landscape. The 'Great Wall of Todmorden' - a massive blue brick retaining wall supporting the adjoining railway's former goods yard - precedes the town itself, a glorious northern community couched under Stoodley Pike, a monument (commemorating both Waterloo and the Crimean War) which sits like a rocket about to be launched on the moors in celebration. By gum they don't make canals like this anymore! Lock 19 is curious in that its tail gate is a guillotine structure which looks like it's escaped from the Nene. A Rochdale Canal Co. boundary stone provides a talking point, as does one

of the Rochdale Canal Society's mile posts, erected - so far - solely on the Yorkshire side of the canal. This one tells you that 'Tod' lies 10 miles from Sowerby Bridge, but doesn't let on how endlessly and provocatively entertaining those ten miles are.

Having had the Roch for company for much of its western ascent to the summit, east of Todmorden the canal has the Calder to converse with. This Calder - not to be confused with the other one which also rises (along with the Irwell) on the moors to the north-west of Todmorden - is an important tributary of the River Aire and, indeed, becomes semi-navigable as a component part of the Calder & Hebble Navigation downstream of Brighouse (Map 35). Not that you'd guess that yet, as, often flowing swiftly in close proximity to the canal towpath, it adds to a watery sense of well-being flowing between the high shoulders of its vertiginous valley. The steepness of the valley sides curtails most views of the moorlands on either flank. Weather-splattered wildernesses of peat bogs and moss are up there unseen, shadowing your progress along the valley's gutter-like floor. But abseiling down to cross the canal by Callis Lock (No.13), the Pennine Way rubber-stamps its seal of approval on your journey, and you may find yourself passing the time of day with travellers of a different sort of gait altogether.

golf course

Todmorden

Dobroyd Castle

30A

30B 24-22

31 21

warehouse

31A

32

mp 25

30

Gauxholme

Town Centre

Lidl

30 19

20 18

Morrisons

Baltimore Marina

29

28

70'

17

27 26

Stoodley Pike

for details of facilities at Todmorden turn to page 76

⚠ CRT facilities key required at Lock 19.

Castle Street

P

16 70'

25 15

Canoeing !

Pickwell & Arnold

24

23 14

70'

Eastwood (csd 1951)

66

22

13 21

PW

32

PW

mp

Eastwood

footpath to Stoodley Pike

N

75

Walsden Map 30

Walsden snuggles cosily in its steep sided valley out of the worst that the moorland winds would otherwise throw at it. It's a mid Pennine community in microcosm, with textile mills and dyeworks (and a strange occupation called 'pickermaking' from Argentinian buffalo hide) taking advantage of the fast-flowing water supply, and with the scars of former quarrying on the slopes. They even dug coal from primitive shafts up on the tops, where the miners rubbed shoulders with hill farmers. The nomenclature resonates with the harshness of life: Top o' th' Rough, Rake End, Jail Hole, Foul Clough Road, Thorns Greece and Pot Oven. It's like another language. And perhaps it was, after all, 'Walsden' is said to derive from "Valley of the Welshmen."

Eating & Drinking

GRANDMA POLLARD'S - adjacent Hollings Lock (No.33). Tel: 01706 815769. Brilliant fish & chip shop and cafe in the same family for nearly a century. Featuring home made pies. They work so hard (from 9am) during the week that they don't open Saturday or Sunday, and they shut at 3.30pm Mon & Tue; and 8.00pm Wed-Fri. OL14 6SA

Shopping

Village store adjacent Bridge 33. Post office adjacent Bridge 35. Gordon Rigg's garden centre and mill shop (with cafe) are about ten minutes walk along the A6033 from the centre of the village - Tel: 01706 817722.

Connections

TRAINS - hourly, daily services linking Walsden with useful towpath-walking stages at Littleborough and Rochdale to the south and Todmorden and Hebden Bridge to the north. Tel: 03457 484950.

The Bear

Todmorden Map 31

Incredible, edible 'Tod' aims to become self sufficient in food supply, a laudable ambition one can only trust it achieves: even Prince Charles has been up to see what's going on. 19th Century Todmorden lay half in Lancashire, half in Yorkshire. The Town Hall - topped by sculpted figures representing the commerce of each county - straddled the boundary. It is typical of the town's rich roll call of fine Victorian buildings enhanced by its setting between the valleys of the Calder and its tributary, Walsden Water. Local hero, John Fielden, is remembered by a statue in Centre Vale Park recalling that he was largely responsible for the Ten Hours Act of 1847, a landmark in industrial reforms preventing employees from working more than a ten hour day which seems to have been conveniently forgotten. Crenellated on its wooded hilltop, Fielden's Dobroyd Castle, latterly a Buddhist retreat, enjoys a new lease of life as a young people's activity centre. Talking of young people, 'Tod' provided a compelling backdrop to the 2004 film *My Summer of Love*, which provided an early starring role for Emily Blunt.

Eating & Drinking

THE BEAR - Rochdale Road (adjacent Lock No.19). Tel: 01706 433606. Vegetarian cafe/restaurant located in former premises of Todmorden Industrial & Co-operative store. Open 9am-5pm daily (10am Sun). Charmingly informal atmosphere. OL14 7LA
BLACKBIRD - Water Street. Tel: 01706 813038. Food served 12-3pm and 4-9pm. OL14 5AB
THE VEDAS - Rochdale Road. Tel: 01706 814009. Indian restaurant adjacent Library Lock (No.18). OL14 7LD
WHITE HART - Station Road. Tel: 01706 811760. Wetherspoon open from 8am daily. OL14 7BD
WHITE RABBIT - White Hart Fold. Tel: 01706 817828. *Good Food Guide* listed contemporary British restaurant open for lunch & dinner (5.30pm) Wed-Sat. OL14 7BD.

Shopping

Marvellous indoor market with heaps of butcher stalls (Mon-Sat, early closing Tue). Open market Wed-Sun. Morrisons and Lidl supermarkets. Two secondhand bookshops: Lyall's on Rochdale Road and the Border Bookshop on Halifax Road (opp Lidl). Locally brewed bottled beer from the Barearts Gallery on Rochdale Road (opp Morrisons).

Things to Do

TOURIST INFORMATION - Burnley Road. Tel: 01706 818181. Well stocked with books and leaflets. OL14 7BU

Connections

TRAINS - four Northern trains per hour (Mon-Sat, hourly Sun) connections with Manchester and Leeds (via Halifax or Dewsbury) and hourly to/from Burnley and Blackburn. Tel: 03457 484950.
BUSES - services along the Calder Valley and to/from Burnley. Tel: 0871 200 2233.
TAXIS - JB. Tel: 01706 814494.

HEBDEN BRIDGE and the Rochdale Canal were made for each other. Restoration of the canal appeared to coincide with the town's reinvention as a northern bastion of the 'new age'; though arguably that was no coincidence at all, cause and effect being the inseparable friends that they are. Cosseted by the wooded valley's steep sides, the town stays its hand to the last possible moment from whichever direction you approach, and the delight is all the more effective for that element of surprise. That said, there will be those of you, like us, who would dearly love to have seen the Calder Valley in its 19th century heyday. Charlestown is a case in point. Here the Manchester & Leeds Railway crossed the canal on a bridge similarly ornate as that at Gauxholme (Map 31) in a setting overlooked by Calderside Dyeworks which boasted a chimney 300ft high. The bridge was replaced by the much more utilitarian structure you see today (Bridge 20) during the Second World War and the works demolished in 1960. In 1912 Charlestown was the scene of a

railway crash when an express came off the rails on a tight curve. Four passengers lost their lives, though a peculiar extra casualty was that of the occupant of a coffin being conveyed for burial in Harrogate, a victim of the sinking of the *Titanic* who might justifiably have taken umbrage at being fatally involved in two transport disasters.

Stubbing Locks (11 and 10) lie picturesquely at right angles to sloping streets of stone-built terraces with back alleys between them. Hebble End Mill, which backs onto the canal, is a centre for alternative technology. Black Pit aqueduct carries the canal across the River Calder at its confluence with Hebden Water in the centre of Hebden Bridge. Descending into guide book bathos, everywhere you look there is something to both catch the eye and captivate it. A slender park separates the canal from the river whilst, on the town side, the canal throws off an arm and drydock where Bronte Boats (Haworth lies merely seven - albeit mountainous - miles to the north-east) provide day boats for hire and other boating facilities.

Falling Royd Tunnel (Bridge 14) was built so that the restored canal could pass beneath the A646. During the first third of the 20th Century an electric tramway followed this busy road. Nowadays - M62 notwithstanding - it hosts an almost constant stream

continued overleaf:

Towpath quality tails off east of Hebden Bridge and NCN 66 detours away at Bridge 14. Well-shod walkers will be fine, but a bit of maintenance wouldn't go amiss.

for details of facilities at Hebden Bridge turn to page 78

for details of facilities at Mytholmroyd turn to page 79

continued from page 77:

of heavy Trans-Pennine traffic to which the quiet of the Rochdale Canal provides piquant contrast. Too piquant in many respects, for it has yet to attract the number of boats its resurgence merits. Like the driest of wines, this Rochdale is an acquired taste. Though once acquired, be warned, there is no going back to the sweet, bland canals of your past.

The canal slips almost unnoticed through the poet Ted Hughes' birthplace, Mytholmroyd. The devastating Calder Valley floods of 2012 and 2015 might well have inspired some dark poetry from the former laureate's pen. But poetry would be of scant consolation to the many homes and businesses affected by seemingly increasing winter deluges.

A good deal of the charm of a boating trip along the heavily-locked Rochdale Canal, lies with the opportunity, every now and then, for you to moor and rest your paddle-gear-weary muscles and walk up on to the neighbouring ridges. Up there the world - or at least this part of the mid-Pennines - is your oyster, and the views can be intoxicating.

To the south of the canal your gaze falls upon Brearley Baptist Chapel and its adjoining Manse. The chapel dates from 1875 and echoes the former role that religion played in alleviating the dull routine of mill work. At the river end of Brearley Lane an end terrace property is known as Little Faith, the locality's original Bethel Chapel.

Brearley Locks lie in an attractive setting overlooked by woods and adjacent to an ancient stone bridge spanning the Calder. Halifax Corporation Tramways - reputedly the hilliest system in the country - operated on tracks of 3ft 6ins gauge. Route No.7 ran out as far west as Hebden Bridge between 1902 and 1936. Evocative postcards of trams running alongside the canal are usually obtainable at the TIC.

Hebden Bridge   Map 32

Hebden's eponymous bridge dates from circa 1520 and still straddles Hebden Water which comes bounding down off the moors to the north; too boisterously in July 2012 when the town was seriously flooded, not once but twice. For some businesses twice was too much to bear, but Hebdenites are made of millstone and grit their teeth determinedly. With an economy reliant now on tourism as opposed to the traditional manufacture of fustian (a sort of thick, twilled cloth), there are even New Age overtones which make it appear, at times, like a northern version of Totnes. But in its setting, deep within the wooded folds of the Calder gorge, and in its sturdy, honey-coloured stone buildings, it transcends any tendency to quaintness, whilst there are many fascinating nooks and crannies waiting to be discovered by the diligent explorer. The poetess, Sylvia Plath, is buried at nearby Heptonstall, a picturesque former weaving village just a short (if steep) walk away.

Hebden Bridge

Eating & Drinking

AYA SOPHIA - Bridge Gate. Tel: 01422 845337. Lively Greek restaurant & cocktail bar. HX7 8EX
CROWN FISHERIES - Crown Street. Traditional fish & chips eat-in/take-away. Tel: 01422 842599. HX7 8EH
DRINK - Market Street. Tel: 01422 844366. Bottled beer shop and microbar. HX7 6EU

FOX & GOOSE - Heptonstall Road. Tel: 01422 648052. *Good Beer Guide* listed community co-operative pub, up to six real ales. HX7 6AZ
MOOCH - Market Street. Tel: 01422 846954. Busy, quirky, friendly cafe bar. HX7 6AA
OLD GATE - Market Street. Tel: 01422 843993. Stylish bar/restaurant: breakfast served from 10am; lunches and dinners.*Good Beer Guide* listed. HX7 8JP
OLIVE BRANCH - West End. Tel: 01422 842299. Turkish/Mediterranean restaurant. HX7 8UQ
ORGANIC HOUSE - Market Street. Tel: 01422 843429. Licensed cafe and shop. HX7 6AA
RIM NAM - canalside. Tel: 01422 846888. Thai restaurant overlooking canal basin. HX7 8AD
STUBBING WHARF - King Street (canalside above Stubbing Locks 10 & 11). Tel: 01422 844107. Vibrant, pub, home cooked food, from noon daily. HX7 6LU

Shopping

The individuality of Hebden's shops is one of its strong-points. Look out for Woodhead's prize-winning butcher's shop on St George's Square, Pennine Wine

& Cheese on Bridge Gate and Holt's fruit & fish shop on the corner of New Road and Bridge Gate, and the interesting Green Shop, canalside at Hebble End. Hebden Bridge Mill on St George's Square houses a collection of shops, eating places and crafts. Being Hebden, there are also heaps of galleries and craft shops. We fell for the bubbly allure of the Yorkshire Soap Company's delightful emporium on Market Street, whilst collectors of rare music on cd and vinyl will be unable to resist the nearby Muse Music & Love Cafe. Market Street also boasts a good independent bookshop called the Book Case. There's a Co-op supermarket on Market Street easily accessed from Bridge 18. Natwest, Yorkshire, Barclays and Lloyds banks, retail market on Thursdays, Farmers Market first and third Sundays in the month.

Things to Do
TOURIST INFORMATION - New Road. Tel: 01422 843831. Canalside information centre with permanent exhibition devoted to canals. HX7 8AF
PICTURE HOUSE - New Road. Tel: 01422 842807. Classic cinema with interesting programmes of off-beat and 'art-house' films. HX7 8AD

Connections
TRAINS - frequent Northern Calder Valley service connecting with Sowerby Bridge and Todmorden, good for towpath walks. Also trains to/from Burnley, Blackburn, Preston etc. Tel: 03457 484950. Nice old-fashioned Lancashire & Yorkshire station with coffee shop and 'pop up' bar (Fri-Sun).
BUSES - services throughout Calderdale. Service 500 (operated by Keighley & District) runs hourly (not to say adventurously) over the moors to Haworth for Brontephiles. Tel: 0871 200 2233.
CYCLE HIRE - Blazing Saddles on Market Street. Tel: 01422 844435. HX7 5TT
TAXIS - Hebden Cars. Tel: 01422 845555.

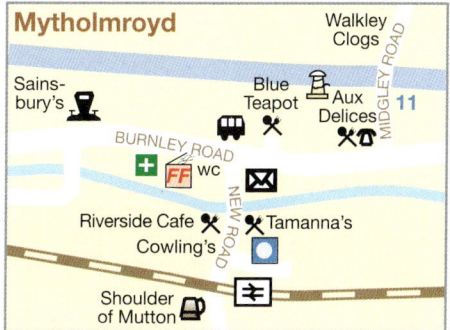

Mytholmroyd   Map 32
Cobbett is said to have slanderously likened Mytholmroyd to the wastes of Nova Scotia. Certainly, around the same time, gangs of lawless 'coiners' operated counterfeit mints in the district. By the time the industrial revolution had kicked-in, however, the village could boast seven cotton mills. Nowadays it is perhaps best known as the birthplace of the poet Ted Hughes who caught his first pike in the Rochdale Canal in the 1930s. A plaque commemorates Hughes' old home on Aspinall Street where he lived between 1930 and 1938. Restored by the Elmet Trust, the house can be booked as a holiday let (4 persons) through Cottages.com - Tel: 0345 498 6900. There are hopes that the three-storey railway station building can be turned into a characterful community centre.

Eating & Drinking
AUX DELICES BISTRO - Burnley Road. Tel: 01422 885564. Bistro and wine bar. Food served Wed-Sat 5-8.30pm and Sun 12-3pm. HX7 5LH
BLUE TEAPOT - Grange Dene Cottage. Tel: 01422 883504. Vegetarian cafe bar and tea room open 9am (10 Sun) to 5pm (4pm Sun). Closed Mon. HX7 5LL

RIVERSIDE CAFE - New Road. Tel: 0775 127 4653. Great for 'Full English' or bacon butties. HX7 5DZ
SHOULDER OF MUTTON - New Road. Tel: 01422 883165. Pub through railway bridge. Open from noon. Food served lunch & evening Mon-Fri and from noon at weekends. Five Yorkshire real ales. HX7 5DZ
TAMANNA'S - New Road. Tel: 01422 885665. Indian restaurant. Lunch & dinner daily. HX7 5DZ

Shopping
Sainsbury's 'Local' on Burnley Road is complemented by Cowlings 'First Choice' greengrocery and delicatessen, the latter being located over the river bridge towards the railway station where, tucked down a side street, you'll also find a launderette. It would be careless of you to pass through Mytholmroyd and not pick up a pair of clogs from Walkley's on Midgley Road (Tel: 01422 885757 - HX7 5LR) they're open Tue-Sat 9.30am-4pm.

Connections
TRAINS - as Hebden Bridge. Tel: 03457 484950.

Luddenden Foot   Map 33
Luddenden Dean flows down from Oxenhope Moor to join the Calder, bridged by a handsome cast iron structure dated 1882.

Eating & Drinking
NEW MOONLIGHT - adjacent Bridge 6. Tel: 01422 886173. Indian restaurant/take-away open from 4.30pm daily. HX2 6AD
AT THE WEAVERS - Burnley Road. Tel: 01422 884523. Wine bar and restaurant. Food served 5-10pm Mon-Fri and from noon at weekends. HX2 6AH

Shopping
Post office with basic groceries.

Connections
BUSES - services 590-4 operate at frequent inteverals to/from Hebden Bridge and Halifax.

33 ROCHDALE/CALDER & HEBBLE Sowerby Bridge 4mls/3lks/2hrs

ARMS linked affectionately with the Calder, the canal counts down the miles to its eastern extremity at Sowerby Bridge on the outskirts of Halifax. Branwell Bronte, the literary sisters' ne'er-do-well brother, was briefly station clerk at Luddenden Foot, and must have been familiar with (if hardly a habitue of) the huge, clock-towered United Reform Church, now converted into apartments, overlooking the canal by Bridge 5. Perhaps the congregation was largely made up of workers from Cooper House Mills and Wood Bottom Dyeworks, two nearby undertakings which have succumbed to the march of time. The railway is a pretty constant companion too. Andrew Martin's literary railway detective, Jim Stringer, was a footplateman with the Lancashire & Yorkshire Railway at Sowerby Bridge engine shed (as vividly described in *The Blackpool Highflyer*) until, one November day in 1905, he inadvertently 'wrecked a locomotive, and a ten-bay engine shed'.

The Calder Valley has been employed as a setting for many television programmes too, most recently Sally Wainwright's police drama starring Sarah Lancashire, *Happy Valley*.

Negotiating a short tunnel, the Rochdale Canal makes an abrupt entry into Sowerby Bridge, where it makes an 'end-on' junction with the Calder & Hebble Navigation. First, though, you must negotiate Tuel Lane Lock, with its intimidating nineteen feet fall, the combined depth of locks 3 and 4, merged when the canal was restored. Two further locks bring you to the Rochdale Canal's junction with the Calder & Hebble Navigation amidst the not inconsiderable splendour of Sowerby Bridge basin.

Does there remain a more visually stimulating canal basin in the country? The last waterborn cargo to reach here was a consignment of paper pulp aboard a keel called *Frugality* in 1955. But, with all the activity of the region's foremost hire base, the setting is still redolent of lost trade. Several handsome stone-built warehouses remain intact which, together with a supporting cast of weighbridge, porters' lodge and agent's house, combine to telling effect. Everywhere you look there are charming details to admire, not least

Such is the depth of Tuel Lane Lock (3) that passage through it is mandatorily overseen by the resident keeper. Between April and October it is open Fri-Mon from 8.30am to 5pm. Tue-Thur and out of season its use must be booked in advance by telephoning 0303 040 4040.

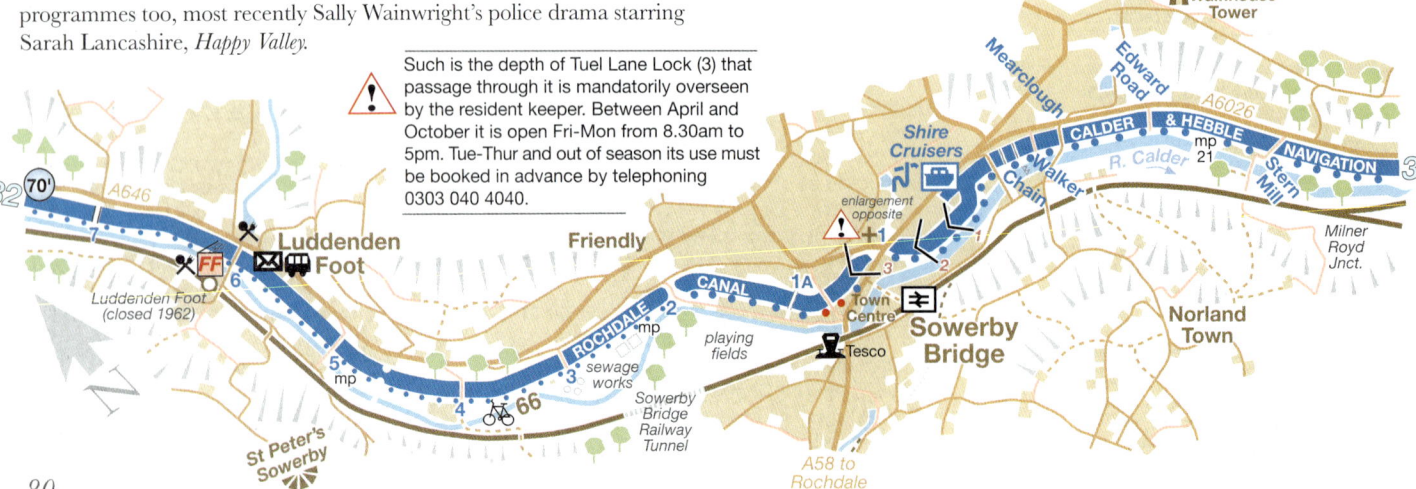

the life-size figures of a boatman and boy putting pressure to a lock beam which welcome folk into the basin off Wharf Street.

The Calder & Hebble Navigation

A change of canal invariably evokes a change of atmosphere whatever the topographical context. Journeying east out of Sowerby Bridge this is immediately true. Not that your eye is necessarily glued to the canal, for it is difficult to draw it away from the sheer vertical impudence of the Wainhouse Tower. Completed in 1875 and thought to be intended as a chimney for a dyeworks, it became a viewing tower instead. Open to the public on selected dates, 369 steps will take you to the Lower Balcony from where, allegedly, it is possible to see Blackpool on a clear day.

Blackpool! Give us Calderdale any day, and you too should by rights be firmly under its spell. Five over-bridges out from Sowerby Bridge you arrive at the site of Sterne Mill (British Waterways - who in their death throes labelled it 'Bridge 4 Stern Mill' - could obviously neither spell nor count) which belonged to the family of the 18th century novelist, Laurence, author of the rambling classic *Tristram Shandy*. Another literary connection here relates to Wordsworth's poem *Lucy Gray*, which was inspired by the drowning of a young girl in the Calder. At the end of the poem Wordsworth suggests that Lucy Gray is still to be seen: 'smoothly tripping along, singing a solitary song, and whistling in the wind', which, minor details apart, renders her indistinguishable from Pearson himself.

Sowerby Bridge   Map 33

There's nothing *sour* about 'Sorby Bridge', neither in its character nor pronunciation. Furthermore the locals are as warm and as welcoming as the name of one of its suburbs - Friendly! A quaint phenomenon is the sight of boaters jostling with shoppers at the Pelican crossing, incongruously wielding windlasses in place of shopping bags as they make their way between Tuel Lane Lock and the pound below. See if you can't time your arrival to coincide with the annual Rushbearing Festival - an orgy of traditional processions and morris dancing - which consumes the town on the first weekend in September. Take yourself down to the Calder Bridge and watch the canoeists going through their paces. Avert your eyes from the Soviet Bloc-like high-rises, foisted on the town in the Sixties. Admire instead the proud Victorian architecture of its public buildings and the still numerous mills. A magical place indeed!

Eating & Drinking

DELI BELGE - Wharf Street. Tel: 01422 836110. Deli with cafe and mouthwatering food to go. Good line in bottled beer. Mon-Sat 8.30am to 5pm. HX6 2AF

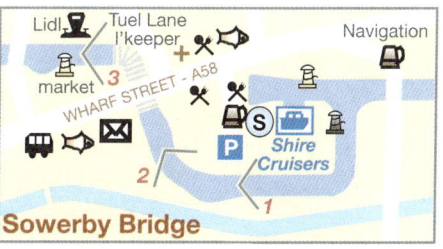

Sowerby Bridge

FIREHOUSE - Town Hall Street. Tel: 01422 832586. *Good Beer Guide* listed establishment where the emphasis is on food (pizza, tapas etc) but with a great choice of real ale. Closed Mondays. HX6 2QD
GIMBALS - Wharf Street. Tel: 01422 839329. Long established 'modern rustic' fine dining. Open Tue-Sat from 6.30pm. Last orders by 9.15pm. HX6 2HF
JUBILEE REFRESHMENT ROOMS - railway station. Tel: 01422 648285. A shared enthusiasm for beer and trains (a winning combination if ever there was one) led the Wright brothers to open this station cafeteria. Hot food up to 2pm ex Sun, brewery memorabilia, and up to six locally sourced real ales. HX6 3AB

NAVIGATION - Chapel Lane. Tel: 01422 316073. Post flood refurbished canalside pub. Closed Mon; food lunch and evening (from 6pm) Tue-Sat and from noon to 5.30pm Sun. Choice of Yorkshire ales. HX6 3LF
SYHIBA - Wharf Street. Tel: 01422 835959. Excellent town centre Indian restaurant. HX6 2AF
TEMUJIN - canal basin. Tel: 01422 835500. Mongolian cooking. Sink your teeth into crocodile, ostrich or kangaroo. HX6 2AG
LA TRADIZIONE - Wharf Street. Tel: 01422 750969. Cosy Italian restaurant. HX6 2EG
VILLAGE - Wharf Street (opposite canal basin). Tel: 01422 831654. Asian cuisine. HX6 2AF

Shopping

Lots of characterful outlets. Lidl handily placed just up the road from Tuel Lane Lock. Large Tesco across the river. Market on Tue, Fri & Sat by Tuel Lane Lock.

Connections

TRAINS - Calder Valley services from station easily reached by footpath and alleyway adjacent to fish & chip shop on Wharf Street. Tel: 03457 484950.
BUSES - services throughout Calderdale. Tel: 0871 200 2233.

Todmorden

Stalybridge

Walsden

13 W

Mossley

Great Wall of 'Tod'

Gauxholme

From
FALLING
19
Miles

Greetland

82

Clegg Hall Mill

Salterhebble

Hebden Bridge

Huddersfield

CLOGGERS KNOLL
UPPERMILL

Registered at Tamworth No.50893

Uppermill

Sowerby Bridge

Stoodley Pike

South Pennine Pleasures

34 CALDER & HEBBLE Salterhebble & Elland 5mls/10lks/3.5hrs

ONCE acquired, a taste for the Calder & Hebble can become all consuming. Atmosphere oozes from every pore of this unjustly undersung waterway: partly because of its unusual gauge (the stumpy locks measure 57ft 6ins x 14ft 2ins, and optically appear very squat indeed); partly because its course lies couched in such a fascinating landscape - half rural, half post-industrial - where interest seldom falters. Too complex to even hint at here, the navigation's history can be traced back to the middle of the 18th century and the West Riding woollen trade's aspirations of building on the success of the earlier Aire & Calder Navigation. Several A-list engineers of that halcyon canal-building era were at one time or another associated with its construction: Smeaton, Brindley, Jessop et al. With the river rarely out of sight, the navigation proceeds through the Calder Valley with locks at fairly regular intervals, echoing a preponderance of weirs on the river itself. Stone mileposts measure the distance from (the faintly amusingly named) Fall Ing where the C & H commences its journey westwards from the A & C at Wakefield. Between Copley and Salterhebble the canal rides above a valley floor littered with sewage farms. A many arched stone viaduct carries the Manchester-Bradford railway across the Calder & Hebble at Copley, a community based on a model mill village of 19th century worker's housing down by the riverside. The village was erected by an altruistic worsted mill owner called Edward Akroyd. The community's impressive French Gothic church, St Stephen, had allegedly to be sited on the far side of the river to appease the Rector of Halifax. No longer used for regular worship, it is cared for now by the admirable Churches Conservation Trust, and contains 'a noteworthy sequence of stained glass by Hardman in the apse'. Bosky footpaths infiltrate the neighbouring woods, Old Rishworthians try to score tries, and there is a splendid little cricket ground boxed in by the railways and the river - do they award *seven* runs if you manage to hit a six over the viaduct?

The trio of locks at Salterhebble are laid out around a dog-leg curve. The top two were orginally a staircase, whilst the bottom lock features an electrically-operated guillotine gate at its tail; though if you've already boated through

See page 100 for details of C&H 'handspikes' and flood locks.

1 Copley - closed 1931
2 Greetland - closed 1962
3 Elland - closed 1962

Map labels:
33 · Copley F'bridge · Copley Viaduct · Copley Lane · cricket ground · weir · Copley · rugby grnd · mp 20 · sewage works · Lister · A629 to Halifax (1 mile) · FF · McD's · Dryclough Junction · S Salterhebble Locks 25ft 0ins · Greetland Junction · Longlees Lock 6ft 6ins · mp19 · garden centre · rugby grnd · Greetland · West Vale · Woodside Mills Lock 6ft 9ins · Elland Bridge · Elland · Town Centre · Morrisons · Valley Mill (con) · mp 18 · Elland Lock 6ft 9ins · Park Nook Lock 7ft 0ins · business park · Knowles Pipeworks · Cromwell Bottom · R. Calder · nature reserve · weir · P · mp 17 · 66 · Crowther · Freemans · A6025 · Cromwell Lock 6ft 0ins · Casa · mp 16 · Camms Mill · Brookfoot · Red Rooster · Brookfoot Lock 5ft 0ins · Ganny Lock 6ft 6ins · 35 · A629 to Huddersfield (4 miles) · N

84

Todmorden, you're entitled to feel blasé about such idiosyncrasies. The gate was manufactured by Ransome & Rapier of Ipswich, a firm perhaps better known - in transport enthusiast circles - for their railway turntables. In the short pound between the middle and lower locks, a small aqueduct spans the Calder & Hebble's junior partner; a shy, retiring sort of chap called Hebble Brook, who springs to life on t' moors above Halifax. A branch canal climbed through fourteen locks in less than two miles to reach the centre of that hill-bounded town. Even as canals go, it was a heavy drinker. And because the local mills already had the valley's water supplies sewn up, water for the branch was pumped (at the rate of a thousand gallons a minute) up from the main line in an Escher-like arrangement of perpetual motion. Opened in 1828, the Halifax Branch was abandoned during the Second World War, and much of its course has subsequently been obliterated, though you get some idea of its topography from the right hand side of a slow-climbing train, or more intimately by walking or cycling the "Hebble Trail". A few hundred yards of the Halifax Branch remains in water at the Salterhebble end, with a Brewers Fayre pub, Premier Inn and McDonald's providing reason enough for a short detour.

Accompanied by the metalled towpath of the Calder Valley Greenway, the Calder & Hebble proceeds through Elland (where it swaps sides) a town once noted for the manufacture of Gannex macs, most famously worn by Prime Minister Harold Wilson. A couple of the locks have long-redundant keeper's cottages alongside them. Two chimneys salute you from the tree-tops above Ash Grove Works, premises of W. T. Knowles, reputedly the last manufacturer in Britain of salt-glazed sanitary pipes and chimney pots.

Elland
Map 34

Characterful small town perched above the Calder. A business park has replaced the mills. The bridge dates back to the 17th century, though with subsequent lengthenings and widenings. Notable businesses include Dobson's boiled sweet factory. Notable buildings include the swimming baths of 1901 (on the outside wall of which a fountain commemorates the demise of the first local man to enlist for the Boer War) and the impossibly quaint Rex Cinema where the organ is still played before the main feature on Saturday nights & recitals on the first Sunday afternoon in the month.

Eating & Drinking
BARGE & BARREL - Park Road. Tel: 01422 371770. *Good Beer Guide* listed canalside pub offering food lunch and evening and up to seven real ales. HX5 9HP
LA CACHETTE - Huddersfield Road (town centre). Tel: 01422 378833. The 'hiding place', a rare French restaurant in these climes. Closed Sun. HX5 9AH
MALT SHOVEL - Briggate. Tel: 01422 373189. Sam Smiths local overlooking Elland Bridge. HX5 9DP

Shopping
Elland provides good shopping facilities (including a Morrisons) and gravity should aid your package-laden progress back to the boat. Dobson's charming sweety shop lies on Southgate.

Connections
BUSES - in the regrettable absence of trains, trams, or trolley-buses, bus services link Elland with the rest of Calderdale. Tel: 0871 200 2233.
TAXIS - Elland Cars. Tel: 01422 327444.

Brighouse
Map 35

Best known for its brass band, Brighouse is an engaging little town with a network of largely traffic-free streets made up of shot-blasted clean Victorian stone buildings. Solely in Pearson's would you glean that Wagner's grand-daughters went to school here.

Eating & Drinking
BLAKELEY'S - Canal Street. Tel: 01484 713907. Fish & chip restaurant and take-away. HD6 1JX
PREGO - Huddersfield Road. Tel: 01484 715566. Canalside Italian (Bridge 18). HD6 1JZ
RICHARD OASTLER - Bethell Street. Tel: 01484 401756. Town centre Wetherspoon housed in converted Methodist chapel. HD6 1JN

Shopping
Canalside Sainsbury's and nearby Tesco apart, the best reason for shopping in Brighouse continues to be Czerwick's wine & cheese (*and* Andrew Jones' pies) shop on Commercial Street, followed closely by Le Gourmet pork pie shop & delicatessen on Bethell Street. Market on Wednesdays and Saturdays.

Things to Do
SMITH ART GALLERY - Halifax Road. Tel: 01422 288065. Enjoyable little gallery bequeathed by local Victorian manufacturer of tweeds and serge. Collection includes an Atkinson Grimshaw. HD6 2AD

Connections
TRAINS - approx hourly links to Huddersfield, Halifax and Leeds (via Dewsbury). Tel: 03457 484950.
BUSES - services E6/7 run to/from Elland; X6/63 to/from Huddersfield and Halifax. Tel: 0871 200 2233.
TAXIS - Woods. Tel: 01484 400800.

IT is at Brighouse that the Calder - which has been beseeching the canal to let it play with it - is finally allowed to join in the fun and games ... and when the river is at its most mercurial, what games they can be! The town's waterfront is dominated by a pair of disued silos, which lend it character, if not a mildly enervating post-industrial despair; and yet, Sugden's Brighouse Mill has found an imaginative new lease of life as a 'climbing gym'. Sugden was a builder of keels before becoming a dusty miller. Wheat was still being brought by barge to the mill when The Beatles first appeared in the hit parade. Anchor Bridge retains its old triangular number plate 'C&H 9'. As does the following bridge, which British Waterways mysteriously renumbered 18!

Brighouse Basin is overlooked by Mill Royd Mill which has been transformed into luxury apartments. On salutary occasions you can't help feeling that the whole of the old West Riding has been turned into luxury apartments. The navigation drops down into the river at the tail of the lower of the two Brighouse Locks. Before it was swept away by floods in 1946, there was a hauling bridge here, enabling boat horses to reach the towing path on the south bank of the river. The Calder injects an element of whimsy - at least where boaters are concerned - down to Anchor Pit Flood Lock, and then again between Kirklees Lower Lock and Cooper Bridge. Despite being spanned by the M62, Kirklees Cut is an isolated length of waterway. Minor details conjour images from the past: the fourteenth milepost from Fall Ing (Wakefield); 100 yard distance posts to the locks; Lancashire & Yorkshire Railway boundary posts; and a curious horizontal pulley wheel which presumably had some function to do with the tow-rope of horse-drawn keels.

Robin Hood and Little John forded the Calder in 1247 at the end of their ill-fated journey to Kirklees Priory, where the hero of Sherwood Forest is said to have been bled to death by the treacherous abbess. At Cooper Bridge the waters of one of the Calder's most significant tributaries, the Colne, make their presence felt amidst an aromatic plethora of sewage works. Good visitor moorings are obtainable in the lock cut, even if you're ultimately headed for the Huddersfield Broad Canal. Collectors of the obscure and enigmatic should stroll down the A62 to its junction with the A644 where the Dumb Steeple, a roadside monument to the efforts of local Luddites to jam the brakes on Progress is to be found.

Map labels:
Sainsbury's Tesco
Clifton Road (csd 1931)
Brighouse
Smith Gallery
Town Centre
Brighouse Locks 12ft 0ins
industrial estate
18
25
weir!
34
industrial estate
Anchor Pit Flood Lock
M62
Bradley Wood Junction
mp 14
R. Calder
Robin Hood's Grave
Kirklees Locks 13ft 3ins
Dumb Steeple
Cooper Bridge (csd 1950)
A644
sewage works
filter beds
monastery
Battyeford Flood Lock
Battyeford Lock 8ft 3ins
to Wakefield
Heaton Lodge Junction
69
Cooper Bridge Lock (C&H) 5ft 3ins
weir!
Cooper Bridge Flood Lock
White Cross
FF
A62 to Huddersfield (3 miles)
Royal & Ancient
PENNINE RING
36
A641

See page 100 for details of C&H 'handspikes' and flood locks.

In the absence of a riverside path, walkers will need to detour from the route the river takes between Brighouse and Anchor Pit Flood Lock. At least they gain some insight into what make's Brighouse's economy tick. Beware erosion on the riverside path between Kirklees Lower Lock and Cooper Bridge.

Note overlap with Map 36

*Flood locks not included

for details of facilities at Brighouse turn back to page 85

36 HUDDERSFIELD CANALS Huddersfield 4mls/12lks/4hrs

WE were not expecting much of the Huddersfield Broad. Consequently we enjoyed it enormously. Firstly, though, we had to hit on its shy, low-profile egress from the Calder & Hebble in the neighbourhood of Cooper Bridge. The old Fearnley Mill chimney serves as a useful landmark, but extreme caution is required from boaters if they are to avoid being drawn towards the boom-protected (and recently rebuilt to the tune of £2 million) weir on the Calder, immediately upstream of which, on the right hand side, lies the entrance lock to the Huddersfield Broad, beneath a slender towpath bridge.

Also known as Sir John Ramsden's Canal (after a local landowner instrumental in its promotion) 'The Broad', opened in 1774, owed its prosperity and longevity to its kinship with the Calder & Hebble more than its association with the Huddersfield 'Narrow' Canal. Commercial trade survived into the 1950s. Overlooked by a stone-built keeper's cottage, Lock 1 is the first of nine which lift the canal up through the Colne Valley to the outskirts of Huddersfield, the best part of four, surprisingly rural miles to the south-west.

Passing under the main Transpennine railway line, and

negotiating Lock 2, the canal slips innocuously through a water treatment works and past the Mamas & Papas factory with its attendant fields of gooseberry bushes. The pound between locks 3 and 4 is framed by abandoned railway viaducts and overlooked by a large modern warehouse occupying the site of a dyeworks which used to get its sulphur from Widnes by barge via the Rochdale Canal. The first viaduct, of fifteen imposing blue brick arches, is a relic of the Midland Railway's ill-timed expansionism in the early years of the 20th century, which was to have opened up the woollen trade and speeded Anglo-Scottish traffic through the West Riding. Their aspirations for a substantial terminus and hotel at Huddersfield did not materialise. The viaduct never carried passenger trains and the track was lifted before the Second World War. A white (or Staffordshire blue-brick) elephant no longer, it has been given a new lease of life as part of the National Cycle Network, and can be reached by footpath from offside

continued overleaf:

⚠ Some of the locks require handcuff keys to unlock the paddle gear. Bridge 17 is electrically operated and requires a CRT facilities key to access the control panel.

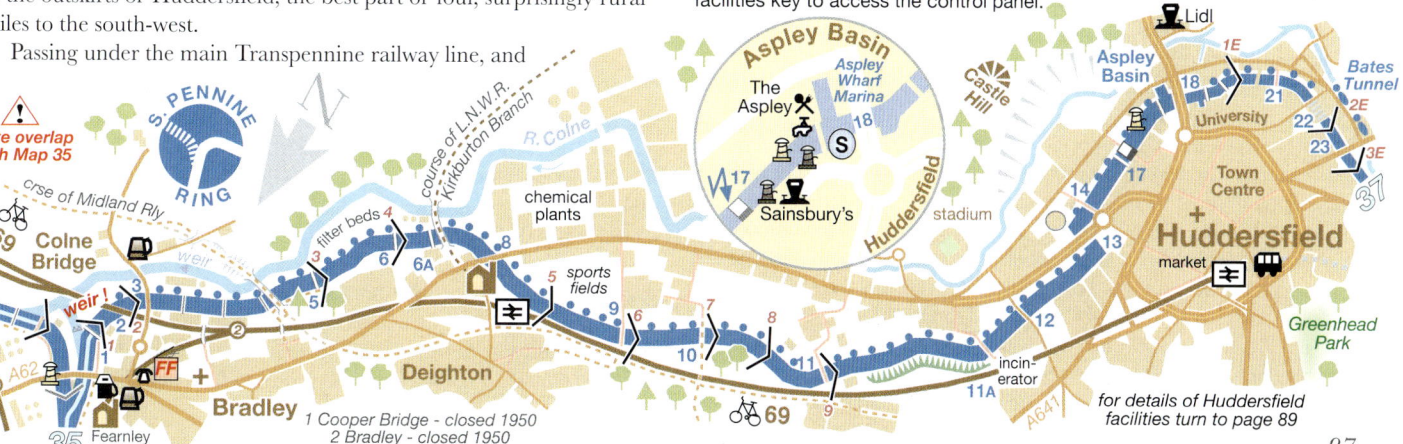

1 Cooper Bridge - closed 1950
2 Bradley - closed 1950
for details of facilities at Bradley turn to page 89

for details of Huddersfield facilities turn to page 89

continued from page 87:

of Lock 3 for a bird's eye view of the neighbourhood. The second viaduct, embellished with ochre brick eyebrows above its blue arches, carried the London & North Western Railway's 'Kirkburton Dick' push & pull locals across the Colne Valley until they steamed into oblivion back in 1930. Those admirable bods at Sustrans have aspirations of converting this viaduct into a cycleway as well ... we look forward with, a warm glow of inevitability, to them doing the M62. An imposing mill stands alongside Bridge 8 at Deighton (pronounced 'Dee-ton'), it used to grind corn, now it deals in hot tubs. Progress personified.

A most attractive pound ensues, with a swarthily wooded railway embankment on one side and an array of sports fields on the other. There are views, to the south-east, of the lofty television mast on Emley Moor. Across Leeds Road Huddersfield Town play football and Huddersfield Giants rugby league at John Smith's Stadium, romantically named - in the modern manner - after the brewery company. What would Herbert Chapman, manager when 'Town' won the league championship three times in a row in the 1920s, have made of the modern vogue for selling a sports ground's name to the highest bidder? The stadium, opened in 1994, is already on its third name.

The high chimney of the municipal incinerator soars above the canal in a manner reminiscent of a similar installation on the Wolverhampton 'Twenty-One'. Less than a mile to go now, and a retail park borders the canal on land once occupied by the town's cattle market and its tram (and later trolleybus) depot. Huddersfield's trolley bus network was one of the most extensive in the country, its hilly district suiting the climbing capabilities of these silently powerful vehicles. So enamoured of this characterful mode of transport were the local populace, a good proportion of them took to the streets to say goodbye when the last of the six-wheeled cream and red fleet returned to the depot for the final time in 1968.

Overlooked by a large gas-holder, 'The Broad' dog-legs sharply under Bridge 13. The gas works had its own railway and the abutments of a bridge which carried the line over the canal remain evident. Led by a flagman, wagons (bearing coal in and coke out) would be hauled by one of the works' Kilmarnock-built Barclay saddle tanks along the neighbouring thoroughfare, giving rise to an ironic sobriquet, 'the Beaumont Street Flyer'.

Two long lost footbridges explain the apparently illogical numbering sequence as the canal proceeds abruptly between bridges 14 to 17. The latter - variously known as Turnbridge, Locomotive or Quay Street - is one of the small wonders of the waterways, designed - if that is not too functional a verb in this circumstance - to lift for the passage of boats. With the Victoria Tower prominent on the Iron Age ramparts of Castle Hill to the south, the canal approaches Aspley Basin. Visitor moorings are available on the towpath side (free for 72 hours courtesy of Aspley Wharf Marina) though they might have been more appropriate on the Sainsbury's supermarket side. At least there is a sense of security for mooring up to explore the never-less-than entertaining town of Huddersfield.

The Huddersfield Narrow Canal

Students, scurrying between lectures, appear oblivious to your enhanced boating credentials as you commence your twenty mile journey across the Pennines at Lock 1E. E, you rapidly realise, stands for 'east' not 'easy'; the paddle gear can be recalcitrant and the intervening pounds thirsty. But then you wouldn't be boating this canal if you were easily daunted.

Initially the Huddersfield Narrow is overlooked by mills refurbished for use by the university. Considerable engineering work was required to restore this section of the canal. Industrial premises had been built over the bed of the canal and in order to by-pass these obstructions it was necessary to lower the canal. Lock 2E was moved west and to reach it the canal passes through a newly built tunnel under the works of Bates & Co Yarn Spinners. The same 'cut & fill' technique was initially adopted to pass beneath Sellars works to reach a new Lock 3E, but this has been opened out as part of the new Waterfront development, and the effect is far less claustrophobic now. Walkers, though, still need to detour between bridges 22 and 23 ... follow the footprints on the map!

Bradley
Maps 35/6

North-eastern suburb of Huddersfield stretching amorphously across the river into Colne Bridge

Eating & Drinking

WHITE CROSS INN - Bradley Road. Tel: 01484 425728. Likeable pub (accessible to west of Bridge 3) serving lunches on weekdays. Tiled Bentleys Yorkshire Brewery initialled insignia in the porch. HD2 1XD.
ROYAL & ANCIENT - Dalton Bank Road. Tel: 01484 425461. Stone-built, Marston's owned country pub and restaurant easily accessible to east of Bridge 3.

Huddersfield
Map 36

Synonymous with Christmas performances of Handel's Messiah, Huddersfield is an enjoyable northern town to visit. Don't rush through! Moor up and explore its dignified streets of Victorian and Edwardian architecture, as evinced by the Corinthian porticoed railway station. Northern pride personified, it stands on one side of a leonine square graced by a statue of the town's famous son, Harold Wilson. Lamentably, the neighbouring George Hotel, birthplace of the peculiarly northern version of the oval ball game in 1895, and scene of Aickman's three baths (see page 92), has bit the dust.

Eating & Drinking

THE ASPLEY - canalside. Tel: 01484 453310. Table-Table branded pub/restaurant, wide range of food. Breakfasts from 6.30am (7am weekends). HD1 6SB
HEAD OF STEAM - St George's Square. Tel: 01484 454533. Lively real ale pub occupying part of Huddersfield's magnificent railway station. Good for food as well as atmosphere. 10 real ales. HD1 1JF
KING'S HEAD - St George's Square. Tel: 01484 511058. *Good Beer Guide* listed rival to above at the opposite end of the station's main westbound platform. Great choice of beer. HD1 1JF

Locomotive Bridge

JULES VERNE - Westgate. Tel: 01484 424078. Refurbished town centre pub serving Carribean and English cuisine daily from noon. HD1 1NP
LALA'S - St George's Square. Tel: 01484 426205. Indian restaurant open daily from 5pm. HD1 1LG
SAN VITO - Kirkgate. Tel: 01484 519699. Italian restaurant over ring-road from Sainsbury's. HD1 1QT
TRATTORIA DOMENICO - Imperial Arcade. Tel: 01484 518588. Town centre Sicilian open daily (ex Sun) from 11.30am. HD1 2BR

Shopping

Excellent shopping centre in streets still laid out in a human, as opposed to car, friendly pattern where the specialist retailers can afford the rents to rub shoulders with the big boys. Splendid open market at the east end of Byram Street open (in various guises) Mon, Tue, Thur & Sat. Indoor market open daily (ex Sun) on Queensgate. Don't miss the Byram Arcade, a Victorian confection of cast iron. Memorable pork pies (pink meat and generous with the jelly) from Mitchell's butchers shop in nearby Station Street.

Things to Do

ART GALLERY - Princess Alexander Walk. Tel: 01484 221964. Strong on 20th Century British art: Bacon, Lowry, Spencer, Atkinson Grimshaw, Henry Moore etc. Tourist information desk too. HD1 2SU

Connections

TRAINS - the railway station is worth seeing even if you have no intention of catching a train! Frequent services to all major Pennine centres including towpather-useful links with Brighouse, Mirfield, Slaithwaite and Marsden. Tel: 03457 484950.
BUSES - service 343 links Huddersfield with Halifax via Elland hourly ex Sun. X63 runs at 10 minute intervals to/from Bradford via Brighouse. Services along Colne Valley corridor, though alas no red and cream six-wheeled trolleybuses. Tel: 0871 200 2233.
TAXIS - Aspley Cars. Tel: 01484 300 030.

Milnsbridge
Map 37

Useful suburban community straddling the Colne and backed by a lofty railway viaduct and imposing mill. Facilities include: pubs, cafes, takeaways; food shops, banks, chemist, butcher, post office and newsagent. Frequent buses to/from Huddersfield.

Linthwaite
Map 37

Colne Valley 'comma' on A62 chiefly notable for characterful, self-brewing Linfit Ales Sair Inn (Tel: 01484 842370 - HD7 5SG), uphill from Lock 17E.

Golcar
Map 37

A 'country mile' uphill from the canal to the north the intrepid hill climber will fall upon the not insignificant delights of the Colne Valley Museum (Tel: 01484 659762 - HD7 4PY) housed in a row of mullion-windowed weaver's cottages. It's open on weekend and bank holiday afternoons, with refreshments available on such occasions.

SHOULDER-CHARGING its way out of the town which gave it its name, the Huddersfield Narrow Canal quickly establishes its unique identity, forming a close alliance with the Colne which is always within earshot if not in view: creaming over weirs, sighing sullenly through man-made channels. It already worked for a living long before the canal came; powering, flushing and cooling for industry, but it was far to mercurial to be made navigable, and the valley had to wait for the canal to be completed in 1811 before it could claim to have a completely reliable mode of transport.

Not an easy canal to negotiate, not an easy canal to build. Overseen by Outram, it was seventeen years in the making: five on the route; twelve on the great tunnel beneath Standedge, of which more on page 92. Being built narrow may have lessened the capital outlay, but it rendered it less able to compete with the railway when it came along, barely forty years later. Soon it came under railway control and traffic inevitably diminished, the last cargo over the summit being recorded in 1905. In 1944 it became a victim of the LMS Railway's infamous Act of Abandonment. All the more remarkable, then, that it should be re-opened to boats in 2001. Ponder this miracle as you plod towards the summit, not doubting that canal restoration can be anything but beneficial in all sorts of unexpected ways.

At Paddock, canal, railway and river form a photogenic whole which finds a strange mirror image at Saddleworth - see Map 40. Here the canal is carried over the Colne on a masonry aqueduct which incorporates Lock 5E. Towering above this unusual arrangement is Sir John Hawkshaw's imposing railway viaduct of 1850, an attractively curvaceous mix of masonry arches and flat latticed ironwork spans. The Colne is bridged again in the vicinity of Golcar, as the canal zigzags through a wooded locality which engenders a typical sense of isolation in an otherwise heavily urbanised valley. Massive textile mills punctuate the valley, none more magnificent than the Titanic Mill which overlooks the wide pound between locks 16E and 17E. Opened in 1912 - the same year that its namesake so infamously sank - and closed sixty-three years later, following a period of decay it has been rejuvenated as apartments and a health spa. One can't help but wonder, however, if the mill - gratuitously sprouting balconies since its make-over - hasn't been emasculated in the process; less Titanic more Titania.

Prone - like us all - to the Second Law of Thermodynamics, the towpath is nevertheless always passable though inclined to muddiness after prolonged rainfall. It improves west of Bridge 40.

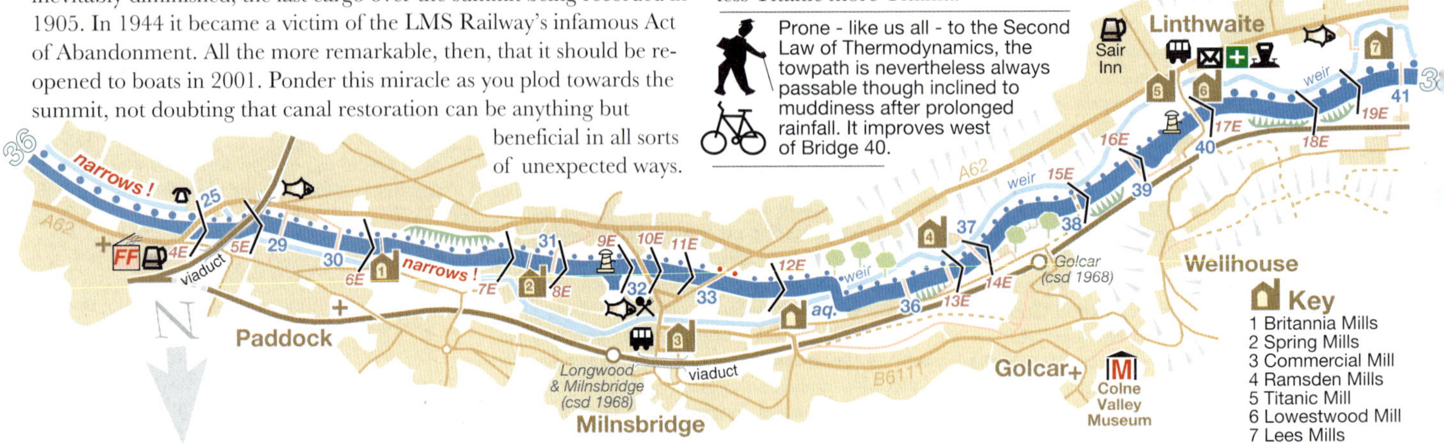

Key
1 Britannia Mills
2 Spring Mills
3 Commercial Mill
4 Ramsden Mills
5 Titanic Mill
6 Lowestwood Mill
7 Lees Mills

Paddock

Milnsbridge

Longwood & Milnsbridge (csd 1968)

Golcar

Colne Valley Museum

Wellhouse

Linthwaite

Sair Inn

Golcar (csd 1968)

for details of Milnsbridge, Linthwaite and Golcar facilities turn back to page 89

ANAL travellers know better than most Britain's ability to reinvent itself topographically from one valley to the next. Few would mistake the Colne Valley for Calderdale, yet they are little more than a lonely patch of moorland apart. Climbing to (and descending from) the highest navigable pound (645ft above sea level) on the British canal system, the Huddersfield Narrow Canal is characterised by the austerity of the valley's high ridges and rough pastures, its mill villages and its stone domestic dwellings, doggedly clinging to the earth with gritted teeth in the face of funnelled gales.

Remarkably transformed by restoration, the canal strides through Slaithwaite with the misplaced nonchalance of a package tourist. The return of the canal here succinctly demonstrates the divers windfalls of canal regeneration; though, more than a decade on, it also illustrates the irritating truth that such achievements need constant loving care if they are not to slip into reverse. Slaithwaite's little shops (along with its 'china cups and virginity') line one side of Carr Lane, the canal the other, backdropped by the grandiloquently-named Globe Worsted Company. If you've still a head on you after passing beneath the confines of Bridge 44, take pleasure from the fact that Upper Mills house both a bakery and a brewery: two of life's shortlisted essentials in mouth-watering propinquity.

Between Slaithwaite and Marsden the canal is at its most bucolic. On foot or afloat, soak up this landscape; drink it all in, from the ponies in the waterside pastures to the stark, horizon-bounding ridges which beg to be climbed. Paths lead tangentially off through bosky glades. Following one across the Colne we came to the A62 where slightly drunken standards, which once supported trolleybus wires, are utilised now for street lighting. Rooks were nesting in the distended guttering of Cellars Clough Mill on the occasion of our latest visit, and a rather officious sign revealed that the premises were in receivership.

continued overleaf:

⚠ !
Lock 24E at Slaithwaite has a guillotine style bottom gate. Access to the hand-wound gear requires a handcuff key. Passage through Standedge Tunnel must be booked in advance with CRT by telephoning 0303 040 4040. Westbound these currently depart from the tunnel entrance on Mondays, Wednesdays and Fridays at 8.30am.

🏠 Key
1 Spa Mill
2 Globe Worsted Co
3 Upper Mills
4 Cellars Clough Mill
5 New Mills

for details of Slaithwaite and Marsden facilities turn to page 93

continued from page 91:

Even in decay, the mill, together with the adjoining Sandhill cottages form an attractive group beside Sparth Reservoir, one of ten built by the canal company to ensure adequate water supplies on such a heavily locked route. The canal actually climbs a total of 438 feet on its eastern side, from which you will extrapolate that the locks average around ten and a half feet per chamber. By the time the westbound traveller reaches Marsden they come thick and fast. Standedge towers above you, brooking no argument, as if you're being pursued and have reached a cul-de-sac. Some dead end! Standedge Tunnel's eastern portal looks like a mousehole in a wainscot. The closer to it you come, the more you have to crane your neck to see the moorland high above you, its tawny skin scarred by wart-like growths of spoil long ago left over from the excavation of the tunnels.

Tunnels plural? Yes, there are *four* of them: the canal tunnel of 1811; the original single bore railway tunnel of 1849; an 1871 duplicate; and a double track railway bore of 1894. The canal tunnel lies below and between the railway tunnels. Robert Aickman named it amongst the Seven Wonders of the Waterways (of which, as you well know, two others are contained in this Companion) and more recently it features (ultimate accolade!) in the *Guinness Book of Records* as the longest (3 miles, 418 yards), deepest (being a maximum of 638 feet beneath the highest point of the moor) and highest, as you already know, in the country. Its construction, given the lack of sophisticated equipment available, is simply astonishing. Travel through it in awe of the boldness of its undertaking, but try also, if you can, to journey over the top and gain a different perspective on the achievement. Disappointingly, the authentic route of Boat Lane, along which horses were led over the top in time honoured canal fashion, has been partially obliterated by subsequent road building programmes, but determined walkers will be able to map out alternative paths. Alternatively a ten minute drive along the A62 will illustrate the tunnel-builders bravery: better still, board a number 184 bus and bag (employing your elbows where necessary) the front seat on the top deck.

Chaperoned by CRT, it takes about an hour and a half to boat through Standedge tunnel today, and it constitutes a great adventure by any standards. The interior is only intermittently lined. At intervals it widens into limestone caverns of great natural beauty: the inland waterways equivalent of the Cheddar Gorge. Four passing places were also provided towards the middle of the tunnel. There's also an intriguing S-bend brought about by a miscalculation as to exactly where the two excavations, dug simultaneously from either end, would meet.

From time to time you come upon linking passages up to the railway tunnels. These 'adits' were used by the railway builders to extract spoil by boat. In the days of steam, dense clouds of railway engine smoke would billow down into the canal tunnel, emerging, in the fullness of time from the Marsden or Diggle portals as though the tunnel were the lair of fiery dragons.

Working boats were 'legged' through Standedge, three and a half hours being the average time to complete a passage. Control of boat movements through the tunnel was apparently entrusted to a boy of twelve, one Thomas Bourne. He must have been good at the job, as he kept the position for a little matter of thirty-seven years, during which he is calculated to have travelled almost a quarter of a million miles backwards and forwards across Standedge Moor, taking charge of the boat horses up to four times a day. Bourne became known as the 'Standedge Admiral' and - goodness knows how - found time to father five daughters and four sons.

In 1824 a steam tug was tried, but only lasted nine years before the leggers took over again. In any case trade through the tunnel was never exactly brisk, and in the second half of the 19th century the railway quickly attracted the bulk of trade. In 1948 - four years after the canal was officially abandoned - a party of IWA luminaries navigated the whole of the Huddersfield Narrow in a wooden cruiser called *Ailsa Craig* and both Robert Aickman and Tom Rolt have described that illustrious journey through Standedge Tunnel in their respective autobiographies *The River Runs Uphill* and *Landscape With Canals*. They were so sooty when they reached Marsden, that they had no alternative but to catch a train down into Huddersfield and have a bath at the George Hotel. Aickman wrote that he actually needed *three* baths, one after the other.

Slaithwaite Map 38

One cannot help but admire the way these Colne Valley communities have reinvented themselves post-industrially. This one even boasts an orchestra! The authentic pronunciation is akin to 'Slough-it' (as in 'come friendly bombs and fall on Slough').

Eating & Drinking

ASHBY'S DELI CAFE - Britannia Road. Tel: 01484 847513. Open 8am-5pm Mon-Fri; 8am-4pm Sat and 9am-3pm Sun. HD7 5HF

CAPTAIN'S TABLE - Carr Lane. Tel: 01484 841068. Fish & chip restaurant (takeaways too) overlooking canal. HD7 5AN

COMMERCIAL - Carr Lane. Tel: 01484 846258. Free house offering up to nine draught beers, not least those from the local Empire Brewery. HD7 5AN

HANDMADE BAKERY CAFE - Upper Mills (canalside betweens locks 23 and 24). Tel: 01484 842175. Open Tue-Sun, 9.30am-4.30pm. Sandwiches, soups, salads, cakes and patisserie in the airy modern premises of a converted mill. HD7 5HA

THE LITTLE BRIDGE - Bridge 44. Tel: 01484 846738. Charming canalside wine bar. HD7 5HF

MONSOON - Britannia Road. Tel: 01484 845818. Tandoori restaurant open evenings daily. HD7 5HG

THE WATERMILL - Station Road. Tel: 01484 845373. HD7 5AW

Shopping

Slaithwaite's array of waterside shops are difficult to resist, and ooze with character: from Blackburn's long established gentlemen's outfitters (where you can obtain a flat cap so as to masquerade as a local) via E. Grange & Son the butcher who makes his own award-winning pies, to the Green Valley Grocer. But our new special favourite is the canalside Handmade Bakery, an admirable not-for-profit, community supported artisan bakery which utilises organic flour and locally-sourced ingedients. Open Tue-Sun from 9.30am-4.30pm. Co-op open 8am to 10pm daily, Lloyds bank with ATM and handy launderette by Bridge 44. Aldi supermarket on Britannia Road.

Connections

TRAINS - hourly, daily Northern local services to/from Huddersfield, Marsden, Stalybridge, Manchester etc. Tel: 03457 484950.

TAXIS - Slawit & Golcar Cars. Tel: 01484 847222.

Marsden Map 38

Quintessential Pennine mill village with an annual jazz festival and associations with the poet Simon Armitage. The superb Mechanics Institute is headquarters of the Mikron Theatre Company who've been travelling the canals dispensing entertainment and inspiration in equal dollops for even longer than we've been producing the *Canal Companions*. It's worth following the Kirklees Way southwards out to Butterley Reservoir (ask directions locally) to view the astonishing spillway, a magical piece of late 19th/early 20th century civil engineering.

Eating & Drinking

THE OLIVE BRANCH - restaurant and accommodation on A62 a mile east of Marsden most easily accessed from Lock 31E and well worth the walk. Tel: 01484 844487. Closed Mon. HD7 6LU

MOZZARELLA'S - Peel Street, village centre. Tel: 01484 845511. Stylish pizzeria under same ownership as The Olive Branch above. Open from 6pm Mon-Thur and progressively earlier into the weekend. Takeaways also available. HD7 6BW

MARSDEN FISHERIES - Peel Street. Tel: 01484 520483. Eat in or t/a fish & chips. HD7 6BW

MARSDEN MOOSE - Market Place. Tel: 01484 843550. Delightful village centre cafe. HD7 6BY

PEEL ONE - Peel Street. Tel: 01484 846800. Smartly furnished modern grill and tapas bar. Food from 5pm weekdays, from noon weekends. HD7 6BR

RIVERHEAD BREWERY TAP - Argyle Street (village centre) Tel: 01484 844324. Atmospheric venue in overall control of the Ossett Brewing Co, but pretty much left to its autonomous devices. Great for food (upstairs restaurant), whilst Riverhead's own distinctive beers derive their euphonious nomenclature from local reservoirs. *Good Beer Guide* listed. HD7 6BR

Shopping

There's something pleasingly 'new age' about Marsden's shops, a quality which dovetails neatly with the more traditional aspects of a Pennine textile community. The Co-op stocks most requisites, but better still, patronise the specialist independents. Nice charity secondhand bookshop called the Cuckoo's Nest.

Things to Do

STANDEDGE VISITOR CENTRE - Waters Road. (Car-parking by station followed by a bracing walk. Tel: 01484 844298. Conversion of former canal workshop into interpretive centre for the restored canal. Refreshments and boat rides into the tunnel. Through trips on selected dates, booking essential. HD7 6NQ

MARSDEN MOOR ESTATE - acres of spectacular National Trust moorland: public access, guided walks and events. Information office located in former railway goods yard adjacent to Marsden station.

MIKRON THEATRE COMPANY - Marsden is the home of this famous canal travelling theatre group. Tel: 01484 843701 for appearance details etc.

Connections

TRAINS - as per Slaithwaite. Tel: 03457 484950.

BUSES - frequent services to/from Huddersfield and Oldham. First's 184 will take you 'over the top', a salutary illustration of the early 19th century tunnellers' achievement. Tel: 0871 200 2233.

STANDEDGE and Saddleworth Moor, unequivocal Pennine landscapes, separated now by the spurious county boundary between West Yorkshire and Greater Manchester, but previously all-embraced by Yorkshire's traditional West Riding. Regular readers will know of (and hopefully align themselves with) our abhorrence of the 1974 boundary changes. But, on this rare occasion, there is something apposite about the current frontier, and it seems somehow apt that the metaphorical customs check between the eastern and western sides of the Pennines should, for boaters at least, occur underground; if only it was still the old Lancashire which met you with daylight at the western portal of Standedge Tunnel.

If the tunnel's eastern portal at Marsden resembles a mousehole, here at Diggle it has the look of a lock-up garage, and there is barely any sense of high ground beyond. The explanation is that this is not the original western entrance to Standedge, in 1894 the tunnel was lengthened by 220 yards to accommodate the new double track railway tunnel. Prior to that the canal essayed a curving course to the east, entering the hillside in the neighbourhood of the Diggle Hotel.

Thirty-two locks, suffixed on this side of Standedge by the letter W (for West, not wizard), carry the Huddersfield Narrow Canal down through the Tame Valley to Ashton-under-Lyne. And barely have you emerged from the tunnel before being confronted with the first one. See how they differ from the rest of the canal's locks, being single-gated at each end and featuring curiously - one might almost say eccentrically - angled paddle gear. They are couched in a typically robust Pennine landscape between the railway and Diggle Brook; heather spills over the moss-grown dry-stone wall which borders the towpath. Down in the valley two big works rear their confident heads. Warth Mill was built by the Co-operative Wholesale Society in 1911. In its time it wove both wool and cotton, but now houses various light industries, including a furniture maker. The other architecturally imposing premises, was Dobcross Iron Works, concerned with the manufacture of textile looms. Latterly a pallet works, redevelopment - and all which that implies - is pending.

⚠ Passage through Standedge Tunnel must be booked in advance with CRT by telephoning 0303 040 4040. Eastbound these currently depart from the tunnel entrance on Mondays, Wednesdays and Fridays at 1pm.

Diggle
Map 39

There seems something comic, faintly music hall, about Diggle, or at least its name; though if you have just spent a number of subterranean hours in Standedge Tunnel you may well be hysterical enough already. The truth, as always, is more prosaic, and this diminutive erstwhile textile village huddles for protection in the folds of moorland which sweep up to the darkest part of the Pennines. Robert Aickman's unnerving story, *The Trains*, had its origins in an anecdote related to him by one of the porters at Diggle's sadly vanished station.

Eating & Drinking
DIGGLE CHIPPY - Huddersfield Road/Wool Lane (access from Bridge 66). Tel: 01457 874398. Quaint fish & chip cabin featuring Hollands Pies from Baxendale. OL3 5PQ

DIGGLE HOTEL - Station Houses, 2 mins walk from Diggle portal. Tel: 01457 872741. Open from noon daily; no food Mondays. OL3 5JZ

GATE INN - Huddersfield Road (village centre, downhill from Bridge 66). Tel: 01457 871164. Refurbished village pub notable for its tiled entrance. Food from noon daily. Accommodation. OL3 5PQ.

GRANDPA GREENE'S - Wool Lane. Tel: 0779 009 2581. Ice cream cabin alongside Lock 31W. OL3 5JT

Shopping
A small post office store forms the sum total of Diggle's shopping facilities, the imposing Co-operative premises on Sam Road, by the tunnel entrance, having been converted into offices.

Connections
BUSES - service 184 links Huddersfield with Manchester hourly (bi-hourly Sun) via Diggle and Uppermill from stops along the road through the village. The exhilarating journey across Standedge to Marsden can be thoroughly recommended for a fresh perspective on the inhospitable setting of Standedge Tunnel. A supplementary service 184 terminates and starts from the village itself (from stops near the tunnel entrance and Diggle Hotel). Tel: 0871 200 2233.

Dobcross
Map 40

One of Saddleworth's constituent villages. Picturesque hilltop location which was effectively used in the1980 movie *Yanks*. Brass band country.

Eating & Drinking
THE NAVIGATION - Wool Road (adjacent Lock 24W). Tel: 01457 872418. *Good Beer Guide* recommended, brass band orientated pub offering food and a variety of beers. Food Mon-Fri 12-3pm and 5-11pm; Sat & Sun from noon. OL3 5NS

LIMEKILN CAFE - Brownhill Countryside Centre (canalside Lock 23W). Tel: 01457 871051. Delightful canalside eatery open 9am-7pm daily. OL3 5PB

Things to Do
BROWNHILL COUNTRYSIDE CENTRE - canalside above Lock 23W. Tel: 0161 770 5888. Information and exhibitions on the Tame Valley's environment. OL3 5PB

Uppermill
Map 40

In the West Riding of Yorkshire until 1974, but now part of the Metropolitan Borough of Oldham in Greater Manchester. Not all the locals were happy for their boundaries to be not fiddled with, and haven't forgotten, let alone forgiven. Chuckling picturesquely through the village, the River Tame shrugs off political treacheries, watering this unexpectedly zestful former textile village which now exudes the atmosphere of an inland resort. On Whit Friday the famous Saddleworth Brass Band Contest (as featured in the film *Brassed Off*) takes place, whilst in August the Morris dancing Rushcart Festival draws in the crowds.

Eating & Drinking
BETTYS - High Street. Tel: 01457 877100. Eat in or takeaway fish & chips open 11.30am-8pm. OL3 6HR

DINNERSTONE - High Street. Tel: 01457 872566. Anglo-Italian restaurant named after an outcrop on Saddleworth Moor. Lunches and dinners daily ex Mon. Takeaways ex Fri & Sat. OL3 6BD

HEI HEI - High Street. Tel: 01457 878487. Eat in or takeaway Chinese. OL3 6HX

NO.73 - High Street. Tel: 0161 236 3656. Vibrant cafe and deli open from 7am daily. OL3 6AP

SHALIMAR - High Street. Tel: 01457 872576. Well-appointed Asian restaurant. Open for dinner daily from 5.30pm (5pm on Suns). OL3 6BD

WAGGON INN - High Street. Tel: 01457 872376. Stone-built inn opposite museum serving Robinsons of Stockport beer. Bar and restaurant food and accommodation. OL3 6HR

Lots of other pubs, fast food outlets and a preponderance of tea rooms.

Shopping
Strung out along High Street are some excellent independent food shops. A secondhand bookshop adds interest to the rash of gift shops. Look out for Buckley's excellent bakery. More practically there's a NatWest bank with cash machine, a post office, chemist, Co-op and Spar food stores.

Things to Do
SADDLEWORTH MUSEUM & GALLERY - High Street. Tel: 01457 874093. Fascinating museum re-opened in 2016 following major refurbishment. Open 1-4pm daily. OL3 6HS

Connections
BUSES - as Diggle plus 350/3/4 to/from Stalybridge and Ashton-under-Lyne. Tel: 0871 200 2233. TAXIS - Street Cars. Tel: 01457 870000.

SUCH is the restricted nature of the Upper Tame Valley that you're not surprised they had to build the Huddersfield as a 'narrow' canal. But, flippancy apart, things are so tight here that when they came to quadruple the railway in 1885, they had no alternative but to lay the extra tracks on the opposite side of the valley. Known as the Micklehurst Loop, it siphoned heavy freight off the main line for almost eighty years, as many as eighty goods trains snorting through the valley day and night.

Coming down from Diggle, the canal encounters a milder Tame Valley, though one given backbone and character by 19th century industrialisation. You fall into a rhythm of lock-working. So much so, that when there's a noticeably longer interval, as occurs between locks 20 and 19, a discomforting sense of withdrawal hovers and your hand strays involuntarily towards your windlass like a production line worker faced with an unexpectedly empty conveyor belt. Away to the east, on a ridge of hills known colloquially as the 'Pots and Pans', an obelisk erected to commemorate the local dead of the Great War, inspired L. S. Lowry's austere

painting *The Landmark*. An outdoor service is held on this 1200ft above sea level summit each Remembrance Sunday.

Headquarters of the enthusiastic Huddersfield Canal Society, the transhipment warehouse at Dobcross remains one of the canal's most historic features. Its location marks the terminus of the canal between 1799 and 1811 while the small matter of Standedge was being considered. Luckily the simple but effective structure has survived years of neglect and now proudly overlooks a waterpoint, auto pump out, Elsan disposal provision and visitor moorings picturesquely sited alongside a converted mill: no irony intended! Nearby stands one of the Huddersfield Narrow's most visually satisfying locations where Saddleworth's curving and imposing railway viaduct straddles the canal at Lock 23W, the latter built into an aqueduct over the River Tame in a similar manner to Lock 5E on the outskirts of Huddersfield.

Straggling communities whose economy was once firmly based on textiles add interest to the canal's progress. Some of them, like Uppermill and Dobcross, have reinvented themselves for use in the twenty-first century world of the service and leisure industries; others, notably Mossley, eke out a post-industrial existence still centred on the local mill, the terraced street, the fried fish shop and the corner shop; all satisfyingly Pearsonesque!

Greenfield Map 40

Former mill village boasting the only railway station left open in Saddleworth.

Eating & Drinking

THE KINGFISHER - canalside Bridge 80. Tel: 01457 872295. New-build Marston's pub offering a wide range of food throughout the day. OL3 7AE

Shopping

Canalside Tesco and nearby post office.

Connections

BUSES - services 350/4 operate to/from Stalybridge and Ashton under Lyne. Tel: 0871 200 2233. TRAINS - hourly, daily local Northern trains to/from Huddersfield via Marsden and Manchester via Stalybridge. Tel: 03457 484950.

Woodend Map 40

Tiny Tame Valley community between Greenfield and Mossley. Calf Lane, which crosses Bridge 85, currently marks the boundary between the Metropolitan Boroughs of Oldham and Tameside, previously Yorkshire and Lancashire.

Eating & Drinking

ASHIANA - Manchester Road. Tel: 01457 832727. Bangladeshi takeaway. OL5 9BB
CHUTNEY MASSALA - Manchester Road. Tel: 01457 838083. Indian restaurant. OL5 9BB
LAL QUILLA - Manchester Road (uphill from Bridge 87). Tel: 01457 839666. Indian restaurant. OL5 9BL
ROACHES LOCK - Manchester Road (adjacent lock 15W). Tel: 01457 837151. Popular stone-built pub alongside an old bridge over the Tame. Open from noon daily, food being served until 9pm. Thwaites ales and guests. Nice beer garden. OL5 9BB
TOLLEMARCHE ARMS - Manchester Road. Tel: 01457 832354. Homely Robinsons of Stockport alternative to above. OL5 9BG

Mossley Map 40

Time was when 'Mozzley' (in the local patois) was perched on the edge of three counties. Though it seems sadly implausible now that Yorkshire, Cheshire and Lancashire ever foregathered here. The town's coat of arms displays both red and white roses, with a sheaf of corn thrown in for Cheshire. Being a border town, woollen and cotton were manufactured. In 1821 the population numbered twelve hundred. Fifty years later it surpassed thirteen thousand. When in Mossley we can never resist patronizing the little fish & chip shop on Staley Road (east of bridges 91/2). Unselfconsciously timeless - the range dates from 1947 - your fish is cooked freshly to order whilst the owner gossips with regulars and sings happily away at his work. On our most recent visit he regaled his customers (without a trace of irony) with *Home on the Range*. He's open Tue-Fri 12-1.30pm and 5-6.30pm.

Eating & Drinking

BRITANNIA INN - Manchester Road (west of Bridge 91). Tel: 01457 838474. *Good Beer Guide* listed Marston's pub near the railway station. Food from noon until 7.30pm (5pm Sun). Additional beers from local Millstone Brewery and others. OL5 9AJ

Shopping

There's a well-stocked convenience store east of Bridge 91/2. For a wider choice ascend westwards to Manchester Road where more shops, including a post office, cluster by the railway station.

Things to Do

HERITAGE CENTRE - Queen Street. Tel: 01457 838608. Open Wed-Fri 2-4pm, Sat 12-2pm. OL5 9AH

Connections

BUSES - services 217/8 run from canalside stops to/from Ashton and Manchester. Tel: 0871 200 2233. TRAINS - as Greenfield.

Stalybridge Map 41

History stalks you in Stalybridge: one senses that the Plug Riots of 1842 are only lately quelled. The Great War's iconic marching song, *It's a Long Way to Tipperary*, was first sung in the Grand Theatre here in 1912. Its composer, Jack Judge is immortalised - suitably accompanied by a 'Tommy' - in an amusing sculpture outside the former market hall, now used for conferences and weddings. Continue along Trinity Street, past the art gallery and post office and across the river upon a handsome bridge (cast at the Albion Ironworks at Miles Platting in 1867) to the police station which was once a school. How relieved the constabulary must be that the main entry isn't under the entrance conspicuously marked 'Girls'.

Eating & Drinking

STATION REFRESHMENT ROOMS - railway station. Tel: 0161 303 0007. Remarkable real ale bar widely celebrated in beer-drinking and train-watching circles. Up to nine real ales usually available along with black peas and other hearty food. SK15 1RF

Shopping

In Melbourne Street (which crosses the tail of Lock 6W) you'll find some characterful local shops; and we were pleased to see The Tripe Shop still flourishing, reputedly the last of its ilk.

Things to Do

ASTLEY CHEETHAM ART GALLERY - Trinity Street. Tel: 0161 343 2878. Worth a visit for the building's interiors alone, though sadly now open only on 1st and 3rd Saturdays each month. SK15 2BN

Connections

BUSES - Tel: 0871 200 2233.
TRAINS - good service to/from Manchester and Huddersfield and local Tame and Colne valley stations. Tel: 03457 484950.

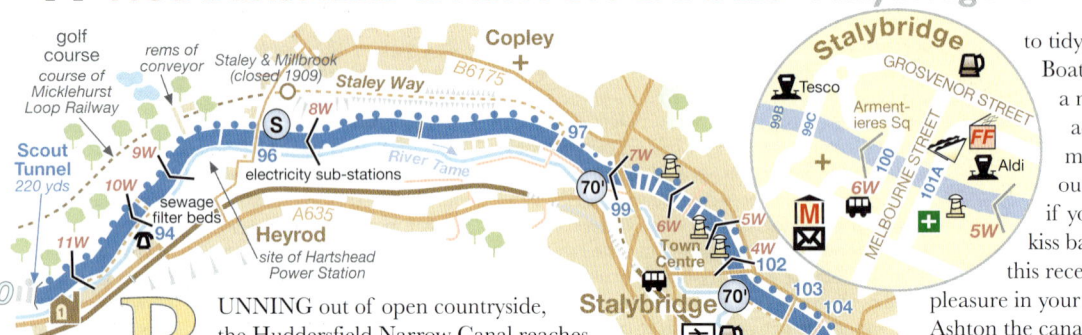

RUNNING out of open countryside, the Huddersfield Narrow Canal reaches the outskirts of Stalybridge and throws in the towel, as if there is only so much beauty a canal can take. Surreal as an art installation, the startling ruin of a half-demolished conveyor looms above the tree-tops below Lock 9W. It used to carry coal from railway sidings to Hartshead Power Station on the west bank of the Tame. We're glad the demolition gang downed tools, and hope its enigmatic, science-fictional remains span the valley for many years to come. Another eerie survivor is Staley & Millbrook station's gargantuan goods shed. CRT provide useful boating services at their maintenance yard above Lock 8W.

Stalybridge embraced the canal's re-opening in 2001 as enthusiastically as Slaithwaite (Map 38). After the 1944 abandonment the line of the canal had been compromised, but happily, with local authority approval and involvement, a bold course was taken and the obstructions simply swept away, resulting in an imaginative regeneration of the town centre, particularly in the vicinity of Armentieres Square; named after Stalybridge's twin-town in France. But now - it gives us no pleasure to relate - apathy seems to have overtaken the town and the canal on its doorstep. Following a serious fire, Armentieres Square resembles its namesake after a bad bout of shelling in the fourteen-eighteen war, and no-one appears in a hurry

to tidy up. A pity after all that effort. Boaters, however, are still regarded as a novelty: pioneers, missionaries; activists in the crusade against the mediocrity which blights so many of our towns. Don't be caught off-guard if you're expected to hug grannies and kiss babies. Play to the gallery and afford this recession-hit populace some vicarious pleasure in your journey. Between Stalybridge and Ashton the canal passes through a corridor of still imposing textile mills, and crossing the River Tame in the process: note the old illegible milepost on the towpath side of the aqueduct. Running slenderly through a stone-walled cutting, the canal passes through its final lock to terminate at Ashton Old Wharf, from where the Ashton Canal leads to Portland Basin and the Cheshire Ring. On arrival boaters are entitled to breath a sigh of relief, but it's unlikely they'll regret exploring the mean, moody and magnificent Huddersfield Narrow Canal.

Key
1 Weir Mill
2 Stamford Mill
3 Premier Mill (Ray)
4 Whitelands Mill
5 Wellington Mill
6 Cavendish Mill

for details of facilities at Stalybridge turn back to page 97, for Ashton-under-Lyne turn back to page 40

This Guide

Pearson's Canal Companions are a long established, independently produced series of guide books devoted to the inland waterways, and designed to appeal equally to boaters, walkers, cyclists and other, less readily pigeon-holed members of society. Considerable pride is taken to make these guides as up to date, accurate, entertaining and inspirational as possible. A good guide book should fulfil three functions: make you want to go; interpret the lie of the land when you're there; and provide a lasting souvenir of your journeys. It is to be hoped that this guide ticks all three boxes, and possibly more besides.

The Maps

There are forty-four numbered maps whose layout is shown by the Route Planner inside the front cover. Maps 1 to 23 cover the main line circuit of the Cheshire Ring commencing at Hardings Wood Junction on the outskirts of Kidsgrove, near Stoke on Trent, and following the route of the ring clockwise via Middlewich, Manchester, Marple and Macclesfield. Map 8A follows the Bridgewater Canal into Runcorn, whilst maps 17A/B illustrate the upper section of the Peak Forest from Marple to Whaley Bridge & Bugsworth. Maps 24/5 show the Bridgewater Canal from Water's Meeting on the outskirts of Manchester to its junction with the Leeds & Liverpool Canal at Leigh. Maps 26 to 41 cover the South Pennine Ring commencing at Piccadilly Basin in Manchester and follwing the route of the ring clockwise via Rochdale, Sowerby Bridge, Huddersfield and Stalybridge to Ashton under Lyne, the last lap of the ring being covered by Map 15 from the Cheshire Ring.

The maps - measured imperially like the waterways they depict, and not being slavishly north-facing - are

INFORMATION

easily read in either direction. Users will thus find most itineraries progressing smoothly and logically from left to right or vice versa. Figures quoted at the top of each map refer to distance per map, locks per map and average cruising time.

An alternative indication of timings from centre to centre can be found on the Route Planner. Obviously, cruising times vary with the nature of your boat and the number of crew at your disposal, so quoted times should be taken only as an estimate. Neither do times

quoted take into account any delays which might occur at lock flights in high season. Walking and cycling times will depend very much on the state of individual sections of towpath and stamina.

The Text

Each map is accompanied by a route commentary placing the waterway in its historic, social and topographical context. As close to each map as is feasible, gazetteer-like entries are given for places passed through, listing, where appropriate, facilities of significance to users of this guide. Every effort is made to ensure these details are as up to date as possible, but - especially where pubs/restaurants are concerned - we suggest you telephone ahead if relying upon an entry to provide you with a meal at any given time.

Walking

The simplest way to explore the inland waterways is on foot along towpaths originally provided so that horses could 'tow' boats. Walking costs little more than the price of shoe leather and you are free to concentrate on the passing scene; something that boaters, with the responsibilities of navigation thrust upon them, are not always at liberty to do. The maps set out to give some idea of the quality of the towpath on any given section of canal. More of an art than a science to be sure, but at least it reflects our personal experiences, and whilst it does vary from area to area, none of it should prove problematical for anyone inured to the vicissitudes of country walking.

We recommend the use of public transport to facilitate 'one-way' itineraries but stress the advisability of checking up-to-date details on the telephone numbers quoted, or on the websites of National Rail Enquiries or Traveline for trains and buses respectively. *continued overleaf:*

continued from page 99:

As reliable as we trust this guide will be, the additional use of a contemporary Ordnance Survey Landranger or Explorer sheet is recommended as they are able to present your chosen route in a broader context. Should you be considering walking the full length of these paths over several consecutive days, the dwindling band of Tourist Information Centres can usually be relied upon to offer accommodation advice.

Cycling

Bicycling along towpaths is an increasingly popular pastime, though one not always equally popular with other waterway users such as boaters, anglers and pedestrians. It is important to remember that you are sharing the towpath with other people out for their own form of enjoyment, and to treat them with the respect and politeness they deserve. A bell is a useful form of diplomacy; failing that, a stentorian cough or the ability to whistle tunefully; light operatic extracts go down very well in our experience. Happily, since the inception of the Canal & River Trust, it is no longer necessary for cyclists to acquire a permit to use the towpath.

Boating

Boating on inland waterways is an established, though relatively small, facet of the UK tourist industry. It is also, increasingly, a chosen lifestyle. There are approximately 30,000 privately owned boats registered on the inland waterways, but in addition to these, a number of firms offer boats for hire. These range from small operators with half a dozen boats to sizeable fleets run by companies with several bases.

Most hire craft have all the creature comforts you are likely to expect. In the excitement of planning a boating holiday you may give scant thought to the

contents of your hire boat, but at the end of a hard day's boating such matters take on more significance, and a well equipped, comfortable boat, large enough to accommodate your crew with something to spare, can make the difference between a good holiday and one which will be shudderingly remembered for the wrong reasons. Traditionally, hire boats are booked by the week or fortnight, though many firms now offer more flexible short breaks or extended weeks. All reputable hire firms give newcomers tuition in boat handling and lock working, and first-timers soon find themselves adapting to the pace of things 'on the cut'.

Navigational Advice

Newcomers, hiring a boat on the inland waterways for the first time, have every right to expect sympathetic and thorough tuition from the company providing their boat. Boat-owners are, by definition, likely to be already adept at navigating; though not necessarily the more demanding waterways of the north. The following, however, may prove useful points of reference.

Locks are part of the charm of inland waterway cruising, but they can be potentially dangerous environments for children, pets and careless adults. Use of them should be methodical and unhurried, whilst special care should be exercised in rain, frost and snow when slippery hazards abound.

In urban areas CRT often attach security appliances to the paddle gear to prevent spurious use of the locks by hooligans. Boaters should ensure that they have at least one - but preferably two or three - 'handcuff' (aka anti-vandal or 'T') keys with which to gain access to the locks thus treated.

The locks on the *Cheshire Ring* are mostly of the narrow variety, the exception being those on the Rochdale Canal in central Manchester which are broad. On the *South Pennine Ring* the locks on the Rochdale, Calder & Hebble, and Huddersfield Broad canals are broad, whilst on the Huddersfield Narrow and Ashton legs of the circuit they are narrow-beam. A peculiarity of the locks on the Calder & Hebble Navigation and Huddersfield Broad is their comparative shortness at 57 feet, a fact which precludes longer craft from using the C&H; though experienced boaters have been known to squeeze a 60ft boat into C&H and HB lock chambers diagonally. Some of the paddle gear on the C&H demands the use of 'handspikes', a length of timber employed in a wrench-like manner and obtainable from boatyards in the vicinity. One further feature of the C&H is the provision of flood locks at the upstream end of navigation cuts which, unless it has been raining heavily, will be 'open' to boats.

Finally, it behoves us all to be on our best behaviour at locks. Remember to exercise a little 'give and take'. The use of foul mouths or fists to decide precedence at locks is one canal tradition not worthy of preservation.

Moveable Bridges are an occasional feature of the routes included in this guide. Some 'swing', some 'lift', some are manually or windlass-operated, some mechanised. Most require either a CRT facilities key and/or 'handcuff' key to facilitate their moving. Always return them to the position you found them in after use unless it is obvious that another boat is approaching to use them.

Mooring on the canals featured in this guide is per usual practice - ie on the towpath side, away from sharp bends, bridge-holes and narrows. Theoretically, you can moor anywhere - as long as the foregoing stipulations are taken into account - but in recent years navigation authorities have signposted designated visitor mooring sights, often with time limitations to dissuade lingering. An open, yellow-tinted bollard symbol on the maps represents such sites; though occasionally we also include this symbol if we feel the location especially lends itself to short term mooring.

Turning points on the canals are known as 'winding holes'; pronounced as the thing which blows because in the old days the wind was expected to do much of the work rather than the boatman. Winding holes capable of taking a full length boat of around seventy foot length are marked where appropriate on the maps. Winding holes capable of turning shorter craft are marked with the approximate length. It is of course possible to turn boats at junctions and at most boatyards, though in the case of the latter it is considered polite to seek permission to do so.

Boating facilities are provided at fairly regular intervals along the inland waterways, and range from a simple water tap or refuse disposal skip, to the provision of sewage disposal, showers and laundry.

Such vital features are also obtainable at boatyards and marinas along with repairs and servicing. An alphabetical list of boatyards appears on pages 102/3. It is worth remembering that facilities on the Rochdale and Huddersfield canals are less prevalent than on the rest of the system, and one should always take the opportunity to fill up with water or dispose of rubbish and sewage when possible.

Closures (or 'stoppages' in canal parlance) traditionally occur on the inland waterways between November and April, during which time most of the heavy maintenance work is undertaken. Occasionally, however, an emergency stoppage, or perhaps water restriction, may be imposed at short notice, closing part of the route you intend to use.

Waterway Authorities

Canal & River Trust

The Canal & River Trust controls the bulk of the inland waterways network. Their Head Office is located at:
First Floor North
Station House
500 Elder Gate
Milton Keynes
MK9 1BB
Tel: 0303 040 4040
www.canalrivertrust.org.uk
The routes contained in this guide are looked after from regional offices at Red Bull (Kidsgrove) and Leeds, both of which can be contacted by telephoning the generic number listed above.

Private Navigations which connect with Canal & River Trust canals covered in this guide are the Bridgewater Canal (Maps 8-14 and 24/5) and Manchester Ship Canal at Pomona Lock (Map 13). The owners of the Bridgewater Canal have a reciprocal

arrangement with the Canal & River Trust allowing boats to pass through without extra charge. Should you need to contact them, however, their head office address is:
Bridgewater Canal Company,
Peel Dome, Trafford Centre,
Manchester M17 8PL.
Tel: 0161 629 8266.

Societies

The Inland Waterways Association was founded in 1946 to campaign for the retention of the canal system. Many routes now open to pleasure boaters may not have been so but for this organisation. Membership details, together with details of the IWA's regional branches, may be obtained from: Inland Waterways Association, Island House, Moor Road, Chesham HP5 1WA. Tel: 01494 783453.
Calder Navigation Society
Huddersfield Canal Society
Inland Waterways Protection Society
Macclesfield Canal Society
Manchester Bolton & Bury Canal Society
Rochdale Canal Society
Trent & Mersey Canal Society
Current membership details of the above groups can be found via the internet or through the Inland Waterways Association.

Acknowledgements

All the usual suspects! Meg Gregory for the signwritten cover; Robin Smithett for additional photography; Karen Tanguy for innumerable classifications of work behind the scenes; and Hawksworth of Uttoxeter for patience and printing. Additional thanks to Nigel and Susan Stevens of Shire Cruisers, Jeff Smith of the Canal & River Trust, and Mike Webb of the Bridgewater Canal Company.

Hire Bases

ABC CANAL BOAT HIRE - Anderton, Trent & Mersey Canal, Map 6. Tel: 0845 126 4098 *www.ukcanalboathire.com* CW9 6AJ

ANDERSEN BOATS - Middlewich, Trent & Mersey Canal, Map 4. Tel: 01606 833668. *www.andersenboats.com* CW10 9QB

BLACK PRINCE NARROWBOATS - Bartington, Trent & Mersey Canal, Map 7. Tel: 01527 575115. *www.black-prince.com* CW8 4QU

BRAIDBAR BOATS - Higher Poynton, Macclesfield Canal, Map 18. Tel: 01625 873471 *www.braidbarboats.co.uk* SK12 1TH

BRIDGEWATER MARINA - Boothstown Basin, Bridgewater Canal, Map 24. Tel: 0161 702 8622. *www.bridgewatermarina.co.uk* M28 1YB

CLAYMOORE NARROWBOATS - Preston Brook, Bridgewater Canal, Map 8. Tel: 01928 717273 *www.claymoore.co.uk* WA4 4BA

HERITAGE NARROW BOATS - Scholar Green, Macclesfield Canal, Map 23. Tel: 01782 785700 *www.heritagenarrowboats.co.uk* ST7 3JZ

MIDDLEWICH NARROWBOATS - Middlewich, Trent & Mersey Canal, Map 4. Tel: 01606 832460 *www.middlewichboats.co.uk* CW10 9BD

SHIRE CRUISERS - Rochdale/Calder & Hebble canals, Map 33. Tel: 01422 832712 *www.shirecruisers.co.uk* HX6 2AG

Day Boats

BAILEY'S TRADING POST - Macclesfield Canal, Map 18. Tel: 01625 872277. *www.baileystradingpost.co.uk* SK12 1TH

BOATING DAYS - Bartington Wharf, Trent & Mersey Canal, Map 7. Tel: 01606 852945 *www.boatingdays.co.uk* CW8 4QU

BOATING DIRECTORY

BOLLINGTON WHARF - Bollington, Macclesfield Canal, Map 19. Tel: 01625 575811. *www.bollington-wharf.com* SK10 5JB

BRIDGEWATER MARINA - Boothstown Basin, Bridgewater Canal, Map 24. Tel: 0161 702 8622. *www.bridgewatermarina.co.uk* M28 1YB

BRONTE BOATS - Hebden Bridge, Rochdale Canal, Map 32. Tel: 01706 815103 *www.bronteboathire.co.uk* HX7 8AD

CLAYMOORE NARROWBOATS - Preston Brook, Bridgewater Canal, Map 8. Tel: 01928 717273 *www.claymoore.co.uk* WA4 4BA

FREEDOM BOATS - Macclesfield, Macclesfield Canal, Map 20. Tel: 01625 420042. SK11 7AW

HERITAGE NARROW BOATS - Scholar Green, Macclesfield Canal, Map 23. Tel: 01782 785700 *www.sherbornewharf.co.uk* ST7 3JZ

PHOENIX DAY BOAT - Whaley Bridge, Peak Forest Canal, Map 17B. Tel: 0775 927 2632. *www.phoenixdayboat.co.uk* SK23 7LS

PORTLAND BASIN MARINA - Ashton-under-Lyne, Peak Forest Canal, Map 16. Tel: 0161 330 3133. SK16 4SQ

THORN MARINE - Stockton Heath, Bridgewater Canal, Map 16. Tel: 01925 265129 *www.thornmarine.co.uk* WA4 6LE

Boatyards

ANDERTON MARINA - Anderton, Trent & Mersey Canal, Map 6. Tel: 01606 79642. CW9 6AJ

ASPLEY WHARF MARINA - Huddersfield, Huddersfield Broad Canal, Map 36. Tel: 01484 514123. HD1 6SD

BALTIMORE MARINA - Todmorden, Rochdale Canal, Map 31. Tel: 01706 818973. OL14 6DA

BARNTON WHARF - Barnton, Trent & Mersey, Map 6. Tel: 01606 783320.

BOLLINGTON WHARF - Bollington, Macclesfield Canal, Map 19. Tel: 01625 575811. SK10 5JB

BLACK PRINCE - Bartington Wharf, Trent & Mersey Canal, Map 7. Tel: 01606 852945. CW8 4QU

BRIDGEWATER MARINA - Boothstown Basin, Worsley, Bridgewater Canal, Map 24. Tel: 0161 7028622. M28 1YB

BRAIDBAR BOAT SERVICES - Higher Poynton, Macclesfield Canal, Map 18. Tel: 01625 873471. SK12 1TH

ELTON MOSS BOAT BUILDERS - Sandbach, Trent & Mersey Canal, Map 3. Tel: 01270 760770. CW11 3PW

FURNESS VALE MARINA - Furness Vale, Peak Forest Canal, Map 17B. Tel: 01663 742971. SK23 7QA

HESFORD MARINE - Lymm, Bridgewater Canal, Map 11. Tel: 01925 754639. WA13 0SW

KERRIDGE DRYDOCK - Kerridge, Macclesfield Canal, Map 19. Tel: 01625 574287. SK10 5AP

KING'S LOCK CHANDLERY - Middlewich, Trent & Mersey Canal, Map 4. Tel: 01606 737564. CW10 0JJ

LYME VIEW MARINA - Wood Lanes, Macclesfield Canal, Map 18. Tel: 01625 858176. SK10 4PH

MACCLESFIELD CANAL CENTRE - Macclesfield, Macclesfield Canal, Map 20. Tel: 01625 420042. SK11 7AW

NEW ISLINGTON MARINA - Manchester, Ashton/Rochdale canals, Maps 15/26. Tel: 0161 839 2999. M4 6BU

NEW MILLS MARINA - New Mills, Peak Forest Canal, Map 17A. Tel: 01663 741310. SK22 3JJ

ORCHARD MARINA - Rudheath, Trent & Mersey Canal, Map 5. Tel: 01606 42082. CW9 7RG

PICKWELL & ARNOLD - Todmorden, Rochdale Canal, Map 31. Tel: 01706 812411. OL14 6DA

PORTLAND BASIN MARINA - Ashton-under-Lyne, Peak Forest Canal, Maps 16/41. Tel: 0161 330 3133. SK16 4SQ

PRESTON BROOK MARINA - Preston Brook, Bridgewater Canal, Map 8. Tel: 01928 719081. WA7 3AF

RENAISSANCE CANAL CARRYING Co. - Oakgrove, Macclesfield Canal, Map 21. Tel: 07791 345004. Fortnightly 'bunkering' service (coal, gas, diesel, chandlery, pump-out etc) Oakgrove-Whaley Bridge; monthly Oakgrove-Preston Brook.

STRETFORD MARINE - Stretford, Bridgewater Canal, Map 13. Tel: 0161 866 8419. M32 0NQ

TRAFALGAR MARINE SERVICES - Newtown, Peak Forest Canal, Map 17A. Tel: 01663 747808. SK22 3HF

UPLANDS BASIN MARINA - Anderton, Trent & Mersey Canal, Map 5. Tel: 01606 782986. CW9 6AJ

WINCHAM WHARF - Wincham Wharf, Lostock Gralam, Trent & Mersey Canal, Map 5. Tel: 01606 44672. CW9 7NT

WORSLEY DRY DOCKS - Worsley, Bridgewater Canal, Map 24. Tel: 0161 793 6767. M28 2WN

Trip Boats

ANDERTON BOAT LIFT - Weaver/Trent & Mersey trips including passage through the lift. Map 5. Tel: 01606 786777. CW9 6FW

CITY CENTRE CRUISES - Manchester & Salford, Bridgewater Canal and Manchester Ship Canal, Maps 13/14. Tel: 0161 902 0222.

HEBDEN BRIDGE CRUISES - Rochdale Canal, Stubbing Wharf, Hebden Bridge, Map 32. Tel: 07966 808717. HX7 6LU

JUDITH MARY - Whaley Bridge, Peak Forest Canal, Map 17B. Tel: 01663 732408. SK23 7LS

MERSEY FERRIES - Manchester Ship Canal cruises between Salford Quays and Liverpool/Seacombe. Tel: 0151 330 1444

SADDLEWORTH CANAL CRUISES - Uppermill, Huddersfield Canal, Map 40. Tel: 0771 118 0496.

Nine More Reasons for Exploring the Canals with Pearsons

9th edition - ISBN 978 0 9562777 4 9

10th edition - ISBN 978 0 9928492 2 1

8th edition - ISBN 978 0 9562777 2 5

1st edition - ISBN 978 0 9928492 1 4

7th edition - ISBN 978 0 9562777 5 6

10th edition - ISBN 978 0 9928492 3 8

8th edition - ISBN 978 0 9562777 9 4

1st edition - ISBN 978 0 9928492 0 7

3rd edition - ISBN 978 0 9562777 6 3

Pearson's Canal Companions are published by Wayzgoose. They are widely available from hire bases, boatyards, canal shops, good bookshops, via Amazon and other internet outlets.